Compound Semiconductor Materials and Devices

Synthesis Lectures on Emerging Engineering Technologies

Editor
Kris Iniewski, *Redlen Technologies, Inc.*

Compound Semiconductor Materials and Devices
Zhaojun Liu, Tongde Huang, Qiang Li, Xing Lu, and Xinbo Zou
2016

Advances in Reflectometric Sensing for Industrial Applications
Andrea Cataldo, Egidio De Benedetto, and Giuseppe Cannazza
2016

Sustaining Moore's Law: Uncertainty Leading to a Certainty of IoT Revolution
Apek Mulay
2015

Compound Semiconductor Materials and Devices

Zhaojun Liu, Tongde Huang, Qiang Li, Xing Lu, and Xinbo Zou

ISBN: 978-3-031-00900-6   paperback
ISBN: 978-3-031-02028-5   ebook

DOI 10.1007/978-3-031-02028-5

A Publication in the Springer series
SYNTHESIS LECTURES ON EMERGING ENGINEERING TECHNOLOGIES

Lecture #3
Series Editor: Kris Iniewski, *Redlen Technologies, Inc.*
Series ISSN
Print 1933-8996   Electronic 1933-9003

Compound Semiconductor Materials and Devices

Zhaojun Liu
Sun Yat-sen University

Tongde Huang
Hong Kong University of Science and Technology

Qiang Li
Hong Kong University of Science and Technology

Xing Lu
Xi'an Jiaotong University

Xinbo Zou
Hong Kong University of Science and Technology

SYNTHESIS LECTURES ON EMERGING ENGINEERING TECHNOLOGIES #3

ABSTRACT

Ever since its invention in the 1980s, the compound semiconductor heterojunction-based high electron mobility transistor (HEMT) has been widely used in radio frequency (RF) applications. This book provides readers with broad coverage on techniques and new trends of HEMT, employing leading compound semiconductors, III-N and III-V materials.

The content includes an overview of GaN HEMT device-scaling technologies and experimental research breakthroughs in fabricating various GaN MOSHEMT transistors. Readers are offered an inspiring example of monolithic integration of HEMT with LEDs, too. The authors compile the most relevant aspects of III-V HEMT, including the current status of state-of-art HEMTs, their possibility of replacing the Si CMOS transistor channel, and growth opportunities of III-V materials on an Si substrate.

With detailed exploration and explanations, the book is a helpful source suitable for anyone learning about and working on compound semiconductor devices.

KEYWORDS

compound semiconductors, III-N materials, light-emitting diodes (LEDs), high electron mobility transistor (HEMT), MOSHEMT, HEMT-LED, device modeling, heterostructures, MOSFET, monolithic integration, epitaxy, current collapse

Contents

CHAPTER 1

Introduction

Since its initial experimental demonstration at Fujitsu Labs in 1980, the compound semiconductor heterojunction-based high electron mobility transistor (HEMT) has been extensively employed in radio frequency (RF) applications, in which low-noise and high-frequency responses are required. Over the past few decades, research on group-III-nitride- and III-V-based HEMTs has received significant attention. Although III-nitrides belong to the III-V families, the stable structure of the group III nitrides, for example, the structure of GaN, is hexagonal wurtzite, and the structures of other III-V materials is cubic zinc-blende.

GaN has attracted substantial amounts of attention as a basis for manufacturing high-speed and high-power electronic devices due to its wide bandgap, high electron saturation velocity, and high critical breakdown field. In particular, GaN-based HEMT, in the form of Al(Ga)N/GaN HEMT, has been regarded as the next-generation RF power amplifier for wireless communication and power converter in utility grid applications. Two main problems associated with GaN HEMTs are gate leakage and DC-RF dispersion (also referred to as knee-voltage walkout), which can be mitigated by implementing a metal insulator semiconductor (MIS) structure and proper surface passivation, respectively. The idea of depositing a layer that simultaneously serves as a gate dielectric layer and as a surface passivation layer leads to an in-depth analysis of *ex situ* Al_2O_3 by atomic layer deposition (ALD) and *in situ* SiN_x by metalorganic chemical vapor deposition (MOCVD).

For high-frequency power applications, several key HEMT quality parameters, including cut-off frequency (f_T), maximum oscillation frequency (f_{max}), and maximum drain current, are linked to a number of important considerations in device design. An overview of HEMT device-scaling technologies that consider barrier thinning, advanced gate fabrication, and S/D resistance reduction is presented.

Due to their tunable, wide, and direct bandgap (0.7–6.2 eV), III-nitride materials have also received significant attention for fabricating light-emitting diodes (LEDs) in the visible light spectrum as general lighting sources. A monolithic integration of GaN-based HEMT and LEDs that can eliminate the parasitic connection inductor and significantly minimize the LED driver size is demonstrated. A number of considerations in the design, growth sequence, and fabrication of this novel integrated system are discussed.

The second section of this chapter presents a detailed overview of zinc-blende III-V based HEMTs. In addition to the diversity of III-V semiconductors that offer superior material selection feasibility for producing heterostructure-based HEMTs, several important types of III-V

HEMT types, including delta-doping HEMTs, "pHEMTs," and "inverted" HEMTs, are described in detail. The status of state-of-the-art V-III HEMTs, which feature high electron mobility in the two-dimensional electron gas (2DEG) layer, are also reviewed, particularly in terms of f_T and $f_{\max}$.

Considering the high electron mobility in III-V HEMTs, the potential for replacing the existing Si CMOS transistor channel with a III-V-based channel has garnered significant interest. A number of key considerations in terms of gate dielectric selection, source/drain engineering, and advanced gate design have been identified and reviewed. To facilitate high-level integration between III-V devices and advanced Si CMOS technology, high-quality III-V thin films on Si, which can be obtained by wafer bonding or heteroepitaxy, are required. The advantages and major challenges in terms of the heteroepitaxy of III-V compounds on Si substrates are discussed. Recent advances in the growth of III-V materials and heterostructures on Si using various buffer layers or nanopatterned growth are comprehensively summarized.

CHAPTER 2

GaN-based HEMTs and MOSHEMTs

2.1 GAN MOSHEMT FOR RF APPLICATIONS

GaN-based HEMT technology has become an important competitor in RF applications because of its outstanding robustness, high Johnson's figure of merit, and low noise figure, which have facilitated GaN becoming an emerging workhorse in the communication market. The last decade witnessed the development of GaN-based devices and monolithic microwave integrated circuits (MMIC), which obtain high efficiency by reducing power consumption, enhancing reliability, and simplifying thermal management. Prior to partially supplanting extensively utilized GaAs- or Si-based devices, two major issues must be addressed by GaN users. Compared with InGaAs-channel HEMTs, GaN-based HEMTs exhibit apparent DC-RF dispersion, which results in a failure to demonstrate the expected RF performance. The proposed dielectric passivation layer is important for recovering the desired performance [1]. This passivation layer primarily functions as compensators to neutralize the trap states. Several types of traps are located in various places throughout the GaN epilayer, as illustrated in Fig. 2.1(b). The traps situated in the buffer always lead to the drain lag, whereas traps on the surface usually cause the gate lag [2]. The generation and recombination process of the traps can modify the charge balance. Various simulations and models have been used to explain this DC-RF dispersion. The common approach to characterizing the trap effects is pulsed I-V, which is extensively adopted to quantitatively and rapidly account for the dispersion. This approach is relatively simple and straightforward and can be directly applied to the walkout of the knee voltage from the pulsed I-V curves, as shown in Fig. 2.1. For a conventional Class-A power amplifier, the output power can be greatly reduced, which significantly limits the gain. This is also applied to other types of amplifiers. The most influential traps are surface traps, which are effectively treated with an extra dielectric passivation layer, such as previously reported SiN_x [3, 4], Al_2O_3 [5, 6], and a SiN_x/Al_2O_3 composite [7, 8]. The barrier and quantum traps are created through the deformation of the epilayer quality near the channel. This creates a hard constraint over the lattice and a thermal mismatch between the barrier and the channel to guarantee a low defect density. The gate leakage current in existing Schottky-gate-diode HEMTs always degrades the noise performance [9, 10] and introduces a potential reliability problem. Based on experience with Si technology, a high-k dielectric is inserted under the gate to impede the leakage current. Consequently, the metal-oxide-semiconductor HEMT (MOSHEMT) or metal-insulator-semiconductor HEMT (MISHEMT) concepts have been developed to reduce

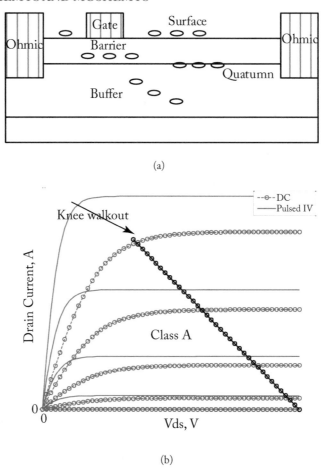

(a)

(b)

Figure 2.1: (a) Locations of potential traps in GaN HEMTs. (b) The effect of the DC-RF dispersion effect on the power efficiency with different knee voltages (consider class A amplifier as an example).

the leakage current in GaN-based HEMTs by several orders of magnitude [11]. The dielectric in the ungated region can simultaneously serve a role in passivation in MOSHEMTs. In this section, we will focus on the GaN MOSHEMTs from the perspective of the gate dielectric for blocking leakage and surface passivation for small RF-DC dispersion.

The gate leakage current in conventional $Al_{0.3}Ga_{0.7}N$ (25 nm)/GaN MOSHEMTs with 10 nm atomic-layer-deposition (ALD) Al_2O_3 as the gate dielectric layer [6] is reduced by a minimum of two orders of magnitude in both the forward and reverse bias after Al_2O_3 deposition. The MOSHEMT has a gate length (L_g) and gate-drain spacing (L_{gd}) of 1 and 2 μm, respectively. The current increases exponentially under forward bias in Schottky contact, whereas the Al_2O_3 effectively blocks the electron injection. No distinct difference between the pulsed and the

DC current is observed for the MOSHEMTs (Fig. 2.2), which demonstrates that the traps at the AlGaN/Al$_2$O$_3$ interface do not deteriorate the pulsed I-V performance. In addition to these con-

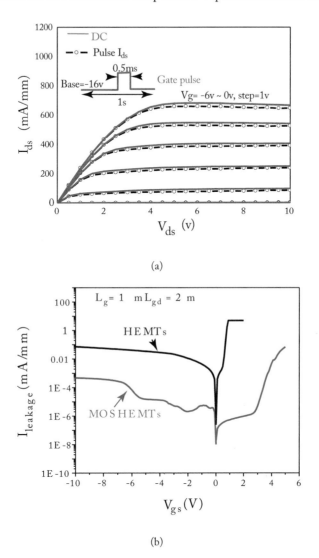

(a)

(b)

Figure 2.2: (a) Pulsed I-V characteristics of AlGaN/GaN HEMTs. A 0.5-ms voltage pulse was applied to the gate with the base voltage maintained at −16 V. (b) Comparison of gate leakage current in HEMTs and MOSHEMTs.

ventional thick-barrier HEMTs, Al$_2$O$_3$ is a passivation candidate for ultra-thin barrier HEMTs such as AlN/GaN [12] and InAlN/GaN [13, 14]. The advantage of a thin barrier is straightforward and preferred in ultra-scaled RF devices to mitigate short-channel effects as scaling down the device gate length. Due to the large tensile strain between the lattice-mismatched GaN and AlN (approximately 2.4%), the growth of a high-quality AlN/GaN interface is challenging. A thin AlN barrier produces surface traps that are very close to the channel. Shinohara et al. demonstrated the outstanding performance of AlN/GaN HEMTs that were grown on SiC substrates by molecular beam epitaxy (MBE) [15], including a cut-off frequency that exceeded 450 GHz and monolithic integrated E/D-mode transistors. However, the performance of AlN/GaN transistors that are grown by metal-organic chemical vapor deposition (MOCVD) is not noteworthy; although MOCVD is a more commonly used technique in the production of GaN-based devices, which is attributed to the higher growth temperature ($>1,000°C$) by MOCVD. Initial AlN/GaN HEMTs were not able to completely pinch off due to the low quality of the AlN barrier layers [16–18]. Medjdoub et al. reported on AlN/GaN HEMTs that were grown by MOCVD on Si substrates with an *in situ* grown SiN_x cap layer for reducing strain relaxation and an increased 2DEG density [19–21]. The SiN_x in the source/drain region must be selectively etched for ohmic contact formation. The reported I_{off} was larger than 10^{-3} mA/mm. Note that the tunneling leakage significantly increases due to the thin barrier, which also renders the gate dielectric indispensable. Huang et al. successfully demonstrated low-leakage AlN/GaN MOSHEMTs that are fabricated using Al_2O_3 as a gate dielectric [12, 22]. The AlN(3 nm)/GaN heterostructure was grown by MOCVD on a 2-inch high-resistance ($\rho > 5000 \, \Omega \cdot cm$) Si (111) substrate. Additional details on the fabrication process and device architecture are provided in [12]. The Al_2O_3 passivation layer can effectively eliminate surface trap states with very large time constants, which trap electrons from channel/gate currents during measurement. To demonstrate the benefits of an ultra-thin AlN/GaN device with Al_2O_3 passivation, enhancement-mode (E-mode) MOSHEMTs were realized with 1.5-nm AlN and source/drain regrowth technology. The cross-sectional transmission electronic microscopy (TEM) image is shown in Fig. 2.3(a), and no obvious relaxation can be seen in the AlN barrier. No boundary is observed between the AlN and GaN because the AlN is pseudomorphically grown on the GaN buffer, which confirms the good quality of the AlN epilayer. Figure 2.3(b) shows a schematic of a cross-section of the E-mode AlN/GaN MOSHEMTs. The regrown n$^+$-GaN technology was employed as the ohmic contact, whose thickness near the mask is approximately 200 nm. The regrowth process is also employed in RF ultra-scaling device fabrication. This topic will be discussed in Section 2.2. The pulsed I-V curves in Fig. 2.4 display a slight but acceptable current degradation in these devices, which is caused by the higher electric field at the drain edge due to the direct connection of the gate to the regrown source/drain. Other techniques, e.g., field plate and large gate-drain spacing, can be utilized to improve performance. The as-fabricated device exhibits a maximum peak G_m value of 509 mS/mm at $V_{ds} = 4$ V. At $V_{gs} = 2$ V, an I_d of approximately 860 mA/mm is attained. These results demonstrate the excellent performance of E-mode GaN HEMTs on Si substrates. By scaling down the gate length, the performance can be improved to be comparable to the performance of GaN HEMTs on SiC substrates.

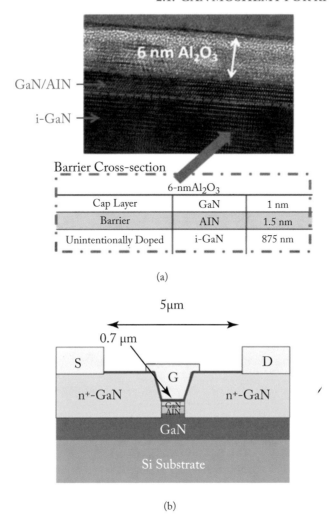

(a)

(b)

Figure 2.3: (a) TEM image of the cross-section of a AlN/GaN heterostructure with an approximately 6-nm Al_2O_3 layer as the gate dielectric and its corresponding schematic; (b) shows the gate length (L_g) of 0.7 μm and the source/drain spacing of 5 μm.

In addition to *ex situ* Al_2O_3 as the gate dielectric, an *in situ* dielectric was also employed in AlN/GaN HEMTs to improve the device performance. The principle advantage of the *in situ* dielectric is the prevention of any external contamination in the following process. The *in situ* SiN_x is deposited in the MOCVD chamber immediately after the GaN HEMTs are grown. Many studies have examined *in situ* SiN_x as the gate dielectric for suppressing gate leakage and the removal of surface traps in AlN/GaN HEMTs [4, 23–25]. Lu et al. developed SiN_x/AlN/GaN

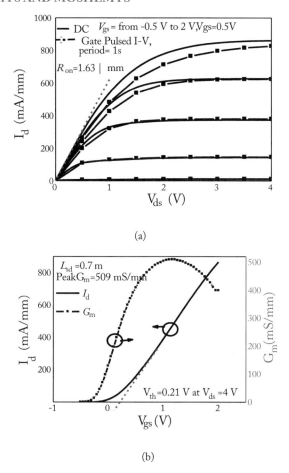

(a)

(b)

Figure 2.4: (a) DC and pulsed $I_d - V_{ds}$ characteristics; (b) transfer characteristics of E-mode AlN/GaN MOSHEMTs with an 6-nm Al_2O_3 layer as the gate dielectric.

MISHEMTs with successfully uniformly grown *in situ* SiN_x of approximately 3 and 7 nm [4, 25]. The *in situ* SiN_x was deposited using silane and ammonia as precursors in the MOCVD chamber at a pressure of 100 mbar and substrate temperature of 1145°C. The SiN_x deposition rate was approximately 2.5 Å/min. X-ray photoelectron spectroscopy (XPS) measurements confirm that the *in situ* SiN_x is also Si-rich, with a N/Si ratio of 1.21, which is similar to the results in [26]. The TEM image in Fig. 2.5(a) shows the uniform coverage of the thin SiN_x with a thickness of 7 nm. Both the SiN_x/AlN/GaN MISHEMTs and circular diodes were fabricated to detect the effects of the SiN_x. Leakage current measurements of the MIS diodes with 3- and 7-nm-thick *in situ* SiN_x layers were performed with the samples biased in the accumulation region. Figure 2.5(b) shows the current density (J_L) versus electric field (E_F) plots of diodes fabricated with various

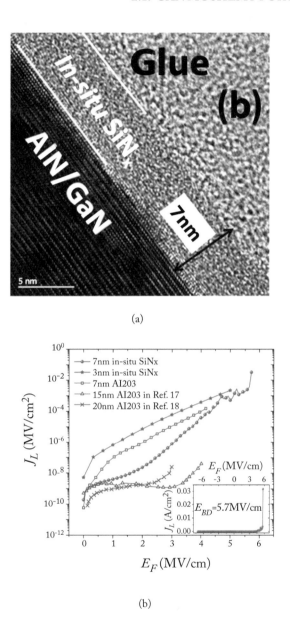

(a)

(b)

Figure 2.5: (a) TEM image of a cross-sectional view of the SiN_x/AlN/GaN structure; (b) leakage current density (J_L) versus electric field (E_F) plots for the *in situ* SiN_x MIS diodes and the Al_2O_3 MOS diodes biased in the accumulation region. The inset diagram shows the breakdown field (E_{BD}) of the 7-nm *in situ* SiN_x.

thicknesses of Al_2O_3 and SiN_x. The direct tunneling current was significantly suppressed as the *in situ* SiN_x film thickness increased from 3 to 7 nm. The reference sample with a 7-nm Al_2O_3 layer shows a higher direct tunneling current than the sample with 7-nm *in situ* SiN_x gate dielectric. The leakage current plots for the thicker Al_2O_3 on conventional AlGaN/GaN heterostructures in [27] and [28] are also included in Fig. 2.5(b) for comparison. At low field, the leakage current of the 7-nm *in situ* SiN_x diode is comparable to the leakage current of the substantially thicker Al_2O_3 MOS diodes. The breakdown field (E_{BD}) for the 7-nm *in situ* SiN_x on AlN/GaN heterostructures is approximately 5.7 MV/cm, as shown in the inset diagram of Fig. 2.5(b). The DC and pulsed I-V curves (Fig. 2.6) exhibit negligible current degradation, which verifies the effectiveness of the *in situ* SiN_x passivation. The capacitance-voltage (C-V) measurements compare the voltage hysteresis (ΔV_{th}) between the Schottky diode and the insulated MOS and MIS diodes. No measurable hysteresis was observed for the Schottky diode, whereas a 0.05 and 0.12 V clockwise hysteresis appeared for the MIS and MOS diodes, respectively. This hysteresis is most likely caused by acceptor-like states in the dielectrics or at the dielectric/barrier interfaces. The capacitance of the 7-nm *in situ* SiN_x diode was calculated to be 1,042 nF/cm^2 by comparing the capacitance of 3- and 7-nm SiN_x MIS diodes. The effective dielectric constant of the *in situ* SiN_x was approximately 8.3, which is higher than the typical value of 7.5 for stoichiometric Si_3N_4. This higher value may be attributed to the higher silicon content of the *in situ* SiN_x [8, 29]. A frequency dependent conductance analysis was performed in the frequency range of 1 kHz to 1 MHz to evaluate the trapping effects in the *in situ* SiN_x/AlN/GaN MIS diodes. Figure 2.7(a) shows measured and modeled parallel conductance (G_P/ω) as a function of the radial frequency (ω) at the vicinity of the threshold voltage (V_{th}) in the MIS diode with 3-nm *in situ* SiN_x. The modeled data are fitted according to (2.1) to extract the trap state density (D_T) and the time constant (τ_T).

$$\frac{G_p}{\omega} = \frac{qD_T}{2\omega\tau_T} \ln\left[1 + (\omega\tau_T)^2\right]. \tag{2.1}$$

A satisfactory fitting was obtained, as shown in Fig. 2.7, and demonstrated the excellent quality of the *in situ* SiN_x gate dielectric. Using the Shockley-Read-Hall statistics in (2.2) the trap state energy (E_T) can be deduced.

$$\tau_T = \frac{1}{v_{th}} \sigma_n N_c \exp\left(\frac{E_T}{kT}\right), \tag{2.2}$$

where $N_c = 4.3 \times 10^{14} \times T^{3/2}$ cm^{-3} is the effective density of the state in the conduction band in GaN, $v_{th} = 2 \times 10^7$ cm s^{-1} is the average thermal velocity of electrons, and $\sigma_n = 1 \times 10^{-14}$ cm^2 is the capture cross-section of the trap states [4]. The trap state density shown in Fig. 2.7(b) for the MIS diodes with 7-nm *in situ* SiN_x layers decreased from approximately 7×10^{12} cm^{-2} eV^{-1} at $E_T = 0.31$ eV to approximately 1.2×10^{12} cm^{-2} eV^{-1} at $E_T = 0.43$ eV. These values are similar to the trap state density of the reference MOS diodes with 7-nm Al_2O_3 layers but are slightly higher than the reported data for conventional Al_2O_3/AlGaN/GaN structures in [30]. The relatively high trap state density is believed to be stemmed from the low quality thin barrier

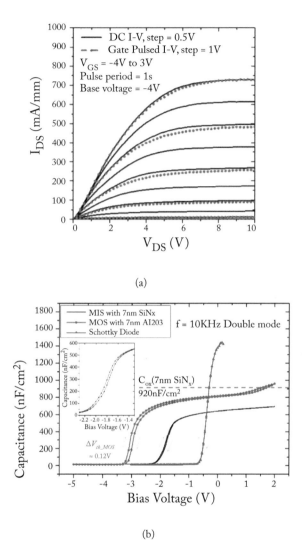

(a)

(b)

Figure 2.6: (a) The pulsed I-V and DC output characteristics. A 500-μs voltage pulse was applied to the gate with a base voltage of -4 V (quiescent bias at pinch-off condition). (b) C-V characteristics of an MIS diode with *in situ* SiN_x, a MOS diode with Al_2O_3 and a Schottky diode. The inset diagram shows a magnification of the hysteresis in the transition region.

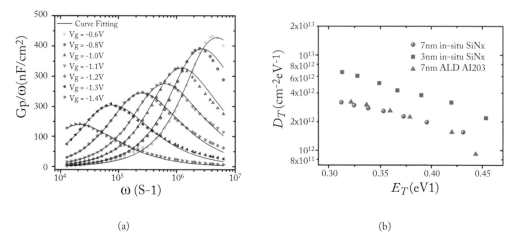

(a)   (b)

Figure 2.7: (a) Frequency dependent parallel conductance as a function of radial frequency for the MIS diode with 3-nm *in situ* SiNx biased with selected gate voltages near V_{th}. (b) Trap state density as a function of energy level depths below the conduction band for the MIS and MOS diodes.

layer of the AlN/GaN heterostructures grown on a Si substrate. The trap state density for the MIS diodes with 3-nm *in situ* SiN$_x$ layers is approximately two times higher than the trap state density of the MIS diodes with 7-nm *in situ* SiN$_x$ layers. This phenomenon has also been observed on Al$_2$O$_3$/AlGaN/GaN structures [30] and can be explained by the different surface passivation effects resulting from the use of SiN$_x$ layers with different thicknesses [25].

The passivation effects of *ex situ* ALD Al$_2$O$_3$ and *in situ* SiN$_x$ on the gate dielectric layer and passivation layer have been reviewed. Traps created at the interface between the semiconductor and the gate dielectric by external air exposure and subsequent processing can deteriorate the device performance and reliability. The *in situ* SiN$_x$ gated devices exhibit smaller voltage hysteresis than that of Al$_2$O$_3$ gated devices in the capacitance-voltage measurement. Recently, SiN$_x$/Al$_2$O$_3$ bilayer gate dielectrics were also investigated to combine the merits of the high band gap of Al$_2$O$_3$ with the advantages of *in situ* SiN$_x$ [23]. This bilayer is considered to be a robust and reliable technique for achieving significantly improved DC and RF characteristics in GaN transistors.

2.2   ULTRA-SCALED-DOWN DEVICE ARCHITECTURE

GaN-based HEMTs have great potential for RF and millimeter-wave power applications [31] due to their unique combination of high electron velocity, large sheet carrier density, and high breakdown field. Since AlGaN/GaN HEMTs were first demonstrated in 1993 [32], remarkable progress has been achieved in terms of improving the transistor performance and reliability, which resulted in state-of-the-art results in terms of cut-off frequency (f_T)/maximum oscillation frequency (f_{max}), and output power density. For example, values of f_T and f_{max} in excess of 300 GHz [15, 33, 34] and output power densities of 1.7 W/mm at 95 GHz [35] and 2.5 W/mm at 40 GHz [36] have been successfully demonstrated in the literature.

f_T of an HEMT is the frequency at which the short-circuit current gain of the device decreases to unity. In the first-order approximation, f_T can be defined as follows:

$$f_T = \frac{g_m}{2\pi(C_{GS} + C_{GD})}.$$

A device with a high f_T will generally function at higher frequencies compared to a device with a lower f_T. A device with higher transconductance (g_m) and lower gate capacitance (C_{GS} and C_{GD}) exhibits a higher f_T. Considering the physical mechanism of HEMT operation, f_T can also be represented by the channel electron drift velocity via the following equation:

$$f_T = \frac{v_{sat}}{2\pi L_G}.$$

The frequency at which an HEMT operates is limited by the electron transit time through the channel. The intrinsic gate delay is usually proportional to the device gate length. A higher electron saturation velocity and smaller gate length result in a higher f_T.

The maximum oscillation frequency (f_{max}) is the highest frequency at which power gain can be obtained from a device. This frequency may be used as an indicator of the ultimate frequency limits of a device. Similar to the f_T value, a high value of f_{max} is desirable if high-frequency operation is desired. For most microwave applications, f_{max} appears to be more useful than f_T because microwave system designers are typically concerned with power gain under matched conditions. f_{max} is defined as the frequency at which the power gain of a device reaches unity. A first-order expression that is frequently used to determine the f_{max} of device can be written as

$$f_{max} = \frac{f_T}{2}\sqrt{\frac{r_{ds}}{R_{gt}}},$$

where r_{ds} is the device output resistance and R_{gt} is the sum of the gate resistance and the channel charging resistance [37].

For high-frequency power applications, f_T, f_{max}, the maximum drain current (I_{DS}) and the off-state breakdown voltage (B_{Voff}) are the critical parameters in terms of the device performance. In addition to the superior material properties, a proper design of the gate configuration and aggressive device scaling are desirable for high-frequency GaN-based HEMTs. As summarized in Fig. 2.8 [15], the use of advanced gate scaling technologies for the reduction in intrinsic and parasitic delay components in GaN HEMTs significantly enhances both f_T and f_{max}. These results were obtained via process maturity in material growth and device fabrication as well as innovative device-scaling technologies. These technologies include heavily doped source/drain (S/D) ohmic contact regrowth [15, 33–35, 38, 39], thin AlN or InAlN barriers [15, 33, 36, 38,

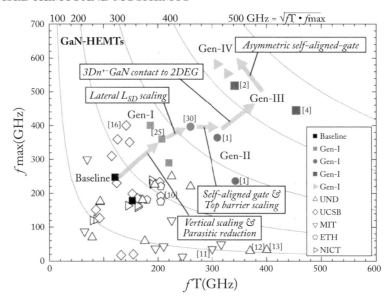

Figure 2.8: Summary of the RF performances for state-of-the-art GaN HEMTs in the literature [15].

40], an advanced T-gate fabrication process [15, 33, 38], and self-aligned structures [34, 39–41, 43].

To facilitate the development of a high-speed, high-efficiency, large power output transistor, small on-state resistance is required to reduce the delay time and dynamic power losses. As the channel dimensions are scaled down, the on-state resistance becomes limited by external resistance sources such as the S/D ohmic contact resistance and the access resistance. Achieving low ohmic contact resistance on a wide bandgap barrier layer of a GaN-based HEMT is difficult compared with traditional Si CMOS technology because the doping level in GaN by ion implantation is significantly lower than the doping level in Si CMOS; the activation efficiency is also very poor in GaN. Multi-layer metals (Ti/Al/Ni/Au) and annealing at high temperatures, typically at approximately 850°C [36, 42], have always been adopted to form alloyed ohmic contact in conventional AlGaN/GaN HEMTs. Such a contact is realized by the diffusion of Ti into AlGaN or GaN to produce n-type doping. The annealing process results in a poor surface morphology and a rough metal boundary, which limits the minimum feature sizes that can be controlled during fabrication. Thus, this alloyed ohmic contact process prohibits the employment of advanced technologies for device scaling such as gate self-alignment. In addition, various spikes that form under the alloyed ohmic contact may cause a high electric field and consequently reduce the breakdown voltage. One emerging alternative is the incorporation of regrown, heavily doped n^+-GaN S/D for ohmic contact formation. Both MOCVD and MBE have been utilized for S/D regrowth [43, 44]. After patterning the dummy gate regrowth mask and recessing the device

S/D region, the sample was re-loaded into an MOCVD or MBE chamber to regrow n$^+$-GaN in the S/D region, as shown in Fig. 2.9. The contact and sheet resistance (R_c and R_{sh}), which are

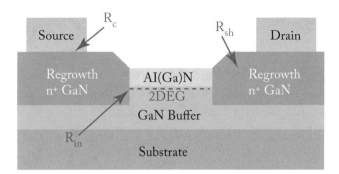

Figure 2.9: Schematic of a cross-section of the S/D regrowth structure.

determined by the regrowth doping level, should be sufficiently small to reduce the device access resistance. In addition, the regrowth interface resistance (R_{in}) associated with the 3D n$^+$-GaN to 2DEG transition region also contributes to the total device on-resistance (Ron). The surface morphology of regrown S/D also serves an important role. Because the gate metal in HEMTs with very short gate lengths require precise positioning, a high step and rough surface will interfere with the alignment accuracy.

As the gate length scales, the remaining parasitic delays in the device should be minimized, and the short channel effects should be considered to facilitate the high-frequency performance of a HEMT. The aspect ratio (the ratio between the gate length and the gate-to-channel separation) is a critical factor that affects the operation of the field effect transistor; a sufficiently high aspect ratio should be achieved to mitigate the short channel effect [37]. However, maintaining the aspect ratio does not guarantee improved performance in GaN-based HEMTs. The reduction in the barrier thickness for a conventional AlGaN/GaN HEMT will result in the degradation of the carrier concentration in the two-dimensional electron gas (2DEG) channel. To overcome these problems, alternative thin barriers, such as lattice-matched InAlN or all-binary AlN, have recently been employed. GaN-based heterostructures with thin wide-band-gap barrier layers can simultaneously achieve a high channel sheet carrier density and reasonable gate control.

Advanced T-gate architectures have been extensively used to obtain high f_T and f_{max} values in many material systems. A T-shaped gate structure can simultaneously achieve a low gate resistance and a low parasitic gate capacitance. A T-gate technology with a highly scaled gate footprint ($L_G < 50$ nm) has also been explored for GaN HEMTs and MISHEMTs. Shinohara et al. demonstrated four scaling generations of T-gate GaN HEMTs, with high f_T and f_{max} values of 450 and 600 GHz, respectively. Figure 2.10 shows cross-sections of this technology [15]. Monolithic integration of E/D-mode GaN HEMTs using ultra-scaled T-gate technology ($L_G < 30$ nm) and high-speed 501 stage ring oscillators have been reported by Schuette et

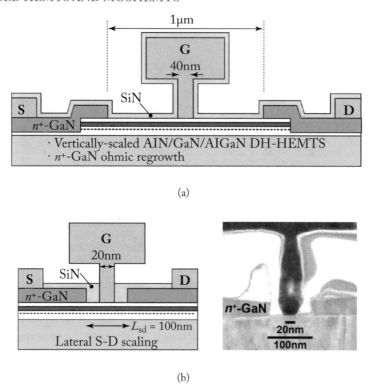

(a)

(b)

Figure 2.10: Cross-sections of four scaling generations of T-gate GaN HEMTs: (a) Generation I. (b) Generation II. *(Continues.)*

al. [33] and Corrion et al. [45], individually. Denninghoff et al. have proposed a high-aspect-ratio T-gate structure to obtain high f_{max}, as shown in Fig. 2.11 [46]. Specially designed 1.1-μm-tall T-gates with a 370-nm-tall stem were employed to simultaneously minimize gate resistance and parasitic gate capacitance. Therefore, a high f_{max} of 351 GHz was achieved for an 80-nm gated device.

Nidhi et al. have developed a gate-first self-aligned structure for N-polar GaN MISHEMTs, as shown in Fig. 2.12. The process involves the deposition of a CVD SiN_x gate dielectric followed by a W/Cr/SiO_2/Cr refractory gate stack, which is defined by e-beam lithography and patterned by dry etching. SiN_x sidewall spacers are formed by blanket deposition and an anisotropic dry etching process to isolate the gate metal from the regrown region. The spacer thickness controls the S/D access distances and the access resistances. The fabrication process and the device architecture are detailed in [34, 39, 47].

Recently, Lu et al. proposed a gate-last self-aligned technology for the fabrication of *in situ* SiN_x/AlN/GaN MISHEMTs [40, 41, 43]. Figure 2.13 illustrates the gate-last self-aligned

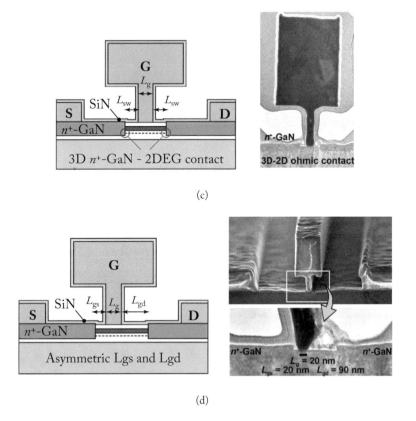

(c)

(d)

Figure 2.10: *(Continued.)* Cross-sections of four scaling generations of T-gate GaN HEMTs: (c) Generation III. (d) Generation IV [15].

fabrication process and device architecture [43]. The gate-last self-aligned process was realized by a dummy gate, which was eventually removed after BCB planarization and replaced with a metal gate. Regrown ohmic contact and low-κ BCB planarization technologies have enabled a reduction in the access resistance and in the parasitic capacitance, which minimizes RC-related delays. The MISHEMTs also featured precisely defined gate-to-regrown-S/D distances by the SiN$_x$ sidewall spacers and an *in situ* grown SiN$_x$ by MOCVD as the gate dielectric. Outstanding device performances, such as a low gate leakage current, a large on/off ratio, and a relatively large $f_T \times L_G$ product, have been achieved for sub-micron gated devices. However, the devices in this first demonstration are less than fully scaled, a limitation brought about by the current photolithography technique. To scale down the device dimensions, an e-beam lithography technique and a single-step dry etch process will be required to define and pattern the SiO$_2$ dummy gate.

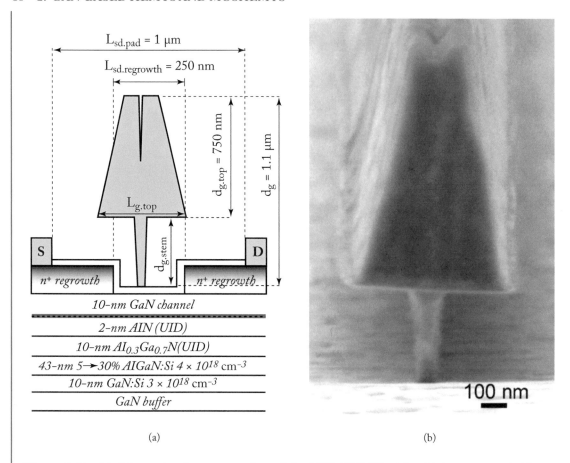

(a)   (b)

Figure 2.11: (a) Schematic of a cross-section of the MIS-HEMT with a high-aspect-ratio T-gate. (b) SEM micrograph of an 80-nm-long, 1.1-μm-tall Ti/Au T-gate with a 370-nm-tall stem formed by electron-beam evaporation and lift-off. The purpose of the tall stem is to reduce parasitic capacitance under the top gate, whereas the purpose of the large top gate is to reduce the gate resistance [46].

Compared with the gate-first self-aligned process, the T-shaped gate formed in this gate-last approach simultaneously enables high f_T and f_{max} values. In addition, this gate-last self-aligned process is less demanding in terms of the gate lithography compared to the conventional T-gate process, which can potentially reduce the process complexity and improve the yield.

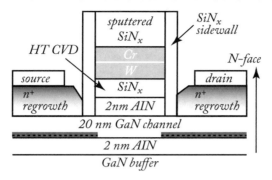

(a)

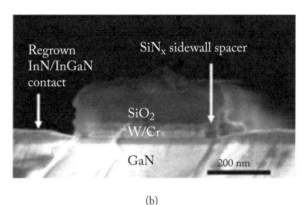

(b)

Figure 2.12: Schematic cross-section (a) and SEM image of the gate-first self-aligned N-polar MISHEMT (b) [39].

2.3 MONOLITHIC INTEGRATION OF GAN HEMT WITH LED

GaN light-emitting diode (LED) represents another semiconductor revolution, which features high power efficiency, and a long lifetime, and is environmentally friendly. Due to the advantages of high efficiency (exceeding 100 lm/W) and long lifespan (maximum of 30,000 hours) over incandescent light bulbs, LED markets have significantly expanded over the last decade. To capture larger segments of the lighting market, the power converters within an LED driver must become more power efficient and smaller. However, LEDs are extremely sensitive to voltage variations and spikes; thus, a driver must ensure a constant power supply. These stringent requirements have promoted researchers to focus their attention on voltage transformers with GaN transistors. GaN transistors can operate at high power levels and high temperature compared with standard Si transistors. The high switching frequency (10 times faster than Si equivalents) can reduce the

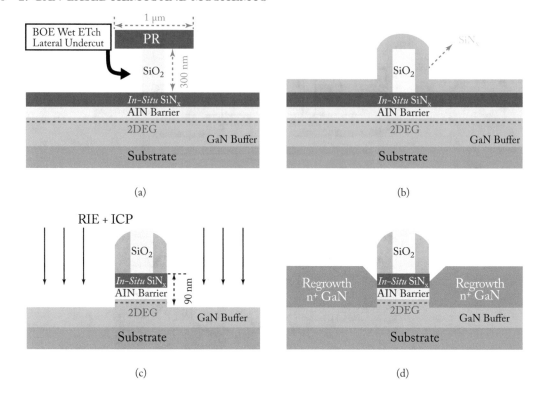

Figure 2.13: Schematic of a cross-section of the gate-last self-aligned fabrication process and device architecture [43].

size of passive components that are built into a driver for energy storage. The figure of merit (*FOM*), which is defined as the product of on-resistance (R_{on}) and the gate charge (Q_g), of a GaN transistor is superior to the *FOM* of a Si transistor [48]. Although numerous studies have investigated high-efficiency GaN HEMT-based converters [49–52], these converters should be wire bonded with an LED chip to supply the required power. This packaged LED-HEMT system would inevitably exhibit parasitic inductance in bonding the GaN switch and the LED, as shown in Fig. 2.14. The parasitic inductor can cause large voltage ringing at the switching node and induce additional power loss. The voltage ringing causes device failure, excessive EMI, instability, and output ripple. This impact may become more significant at high switching speeds. The monolithic integration of GaN-based LEDs and HEMTs is intriguing in terms of addressing this issue and improving the system's reliability with a simplified package. The integration can eliminate the parasitic connection inductor between the LED and the device.

Prior to discussing the integration, the high superiority of GaN to Si power devices is discussed. A typical AlGaN/GaN heterostructure was grown on a 2-inch sapphire substrate by the Axitron MOCVD system [53]. The as-fabricated device without any surface passivation has

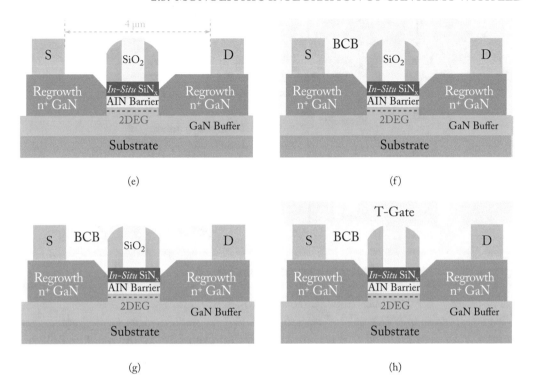

Figure 2.13: Schematic of a cross-section of the gate-last self-aligned fabrication process and device architecture [43].

a threshold voltage of approximately -3 V and exhibits normally on characteristics. As shown in Fig. 2.15, the specific on-resistance is approximately 13 Ω-cm^2 based on the device with a gate-drain distance (L_{gd}) of approximately 15 μm. The transistor has a total gate width (W_g) = 2 mm and a gate length (L_g) = 1.5 μm. No additional gate or source field plate is incorporated to reduce the current dispersion. The reference Si-based power MOSFET is IRF830APbF [54] with an on-state resistance of 1.4 Ω and drain-to-source breakdown voltage of 500 V at $I_d = 250\ \mu$A. The breakdown voltage for the fabricated GaN device displays a maximum voltage of 1,000 V at $I_d = \sim 0.5\ \mu$A. To fairly compare the GaN and Si devices, the gate charges are characterized with an identical measurement setup that excludes additional cooling at room temperature. Note that the GaN is an unpackaged device. The conduction loss is determined by the on-state resistance of the I-V curve, whereas the dynamic loss can be predicted from the gate charge, which is one of the critical parameters for power switching devices. A commercial p-MOSFET was employed as the load current source. The extracted total gate charge (Q_g) for the Si MOSFET is $Q_g = \sim 11$ nC for a gate-to-source voltage (V_{gs}) that ranges from 0–6 V. The measured Q_g results are consistent with that provided in the Si device's datasheet [54]. On the other hand, the total Q_g for the GaN

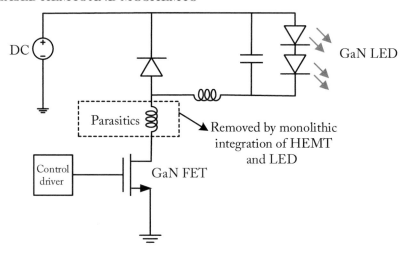

DC

GaN LED

Parasitics

Removed by monolithic
integration of HEMT
and LED

Control
driver

GaN FET

Figure 2.14: Architecture of the LED driver with GaN HEMTs.

device is approximately 0.1 nC for V_{gs} from -5 to 1 V. Both devices were characterized with the same V_{gs} sweep range of 6 V and drain supply voltage of 30 V. The measured $R_{on} \times Q_g$ of GaN (2.95 ΩnC) is significantly reduced compared to that of the Si MOSFETs (15.4 ΩnC), which indicates that GaN can outperform Si in terms of total switching and conduction loss. This confirms the GaN chip area can also be reduced while maintaining the same performance as the Si MOSFET; even this GaN device shows relatively poor performance compared with the state-of-the-art GaN properties [55, 56] due to the high R_{on} for our GaN device.

Based on this analysis, the GaN power device has a great advantage over the Si MOSFET in terms of the power switching converter. The monolithic integration of the GaN device and the LED should be considered. The HEMT-LED integration technology involves two major considerations. The first consideration is the choice between selective area regrowth and dry-etching removal. The second consideration is the large thermal budget required in the MOCVD growth that can degrade the performance of both devices. Recently, researchers from Rensselaer Polytechnic Institute (RPI) discovered a direct epitaxial integration of LED epilayers on top of the layers of an HEMT. The HEMT was subsequently exposed using selective epi removal (SER) with dry etching [57]. However, dry etching will unavoidably introduce plasma damage, and the etching depth is very difficult to control because the AlGaN barrier thickness is only 20 nm under the HEMT. The uniformity of the etching depth across the entire wafer is not guaranteed. To overcome this shortcoming of dry-etching removal, Liu et al. at The Hong Kong University of Science and Technology (HKUST) promoted selective epi growth (SEG) for the integration of an HEMT with an LED [58, 59]. The AlGaN/GaN heterostructure was grown on the top of a commercial LED epitaxial layer on a sapphire substrate. From bottom to top, the epitaxial HEMT structure consisted of a 170-nm undoped GaN layer, a 55-nm AlGaN layer with Al com-

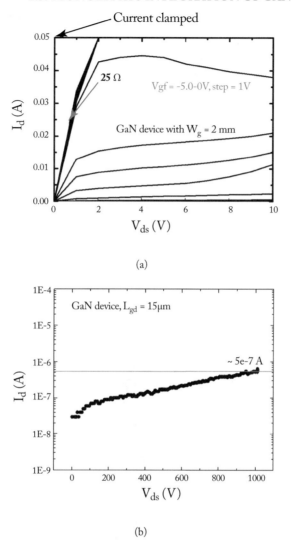

(a)

(b)

Figure 2.15: (a) I-V characteristics of the as-fabricated GaN power device with $W_g = 2$ mm; the maximum on-state current is clamped by the measurement setting. (b) Off-state characteristics of the device at $Vg = -5$ V for a device with $L_{gd} = 15$ μm.

position of 15%, a 150 nm undoped GaN channel layer, a 1-nm AlN spacer, and a 20-nm AlGaN barrier layer. Prior to HEMT growth, 500-nm SiO_2 was deposited on the LED epi-structure and patterned by buffered oxide etchant (BOE) as a growth mask. Considering the LED-epi underneath, the growth temperature of the HEMT-epi was 1,045°C, which was 100°C lower than our normal HEMT growth temperature to prevent the degradation of the quantum well in the LED at the expense of a relatively poor buffer quality of the HEMT. The lower growth temperature may be responsible for the large leakage current of the HEMT devices. Additional growth optimization will be beneficial for low leakage. The SEG and SER approaches are compared in [59] and their schematics are shown in Fig. 2.16. To investigate the effect of two approaches on LED

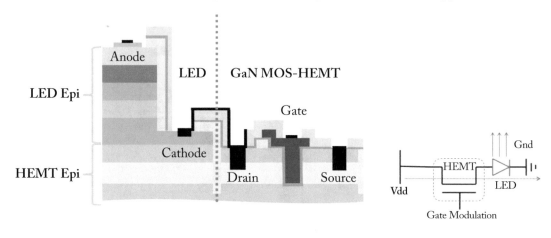

(a)

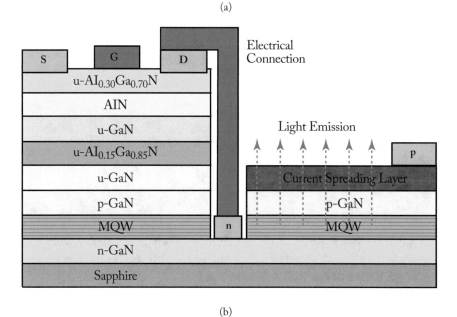

(b)

Figure 2.16: Schematic of a cross-section of the monolithically integrated GaN LED and GaN HEMT using the SER (a) and SEG (b) approaches.

characteristics, a reference LED sample was also fabricated to monitor the process. The LEDs fabricated by SER (Fig. 2.17(a)) and SEG (Fig. 2.17(b)) exhibited a forward voltage (V_F) of 17.1 and 3.7 V, respectively. The reference LED sample showed $V_F = 3.1$ V, which confirmed that

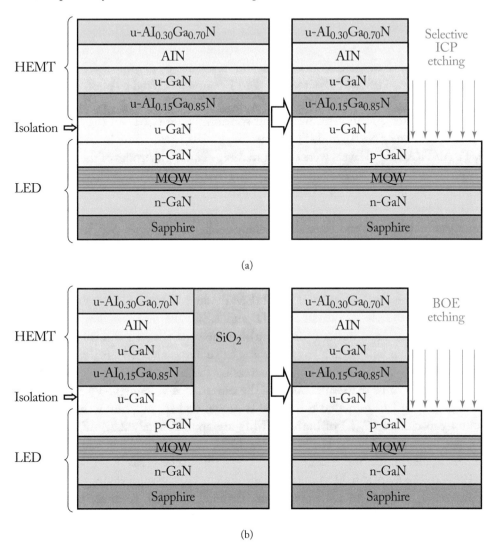

Figure 2.17: Schematic of a cross-section of the (a) SER and (b) SEG approaches.

LEDs by SER are highly resistive and can cause poor power efficiency. As a result, SEG is the preferred approach for the monolithic integration. The reason for this significantly increased V_F is that a nitrogen-deficient surface was produced due to the exposure of the p-GaN to the plasma etching, which introduced donor states with concentrations above 1×10^{20} cm^{-3} in the surface layers. These donor states can cause a significant reduction in the hole concentration or even type conversion of the surface layers [59]. SER was also utilized by researchers at RPI to fabricate integrated HEMTs and LEDs [57]. The V_F of the fabricated LED is significantly higher than 4 V. Therefore, the following paragraphs primarily focus on the SEG technique due to its superiority.

The HEMT fabrication process in SEG began with the complete removal of the SiO_2 growth mask by BOE after epi-growth. Mesa regions of the HEMT were created using Cl_2-based ICP selective etching with the LED regions protected by photoresist. Next, mesa regions of the LED were created using ICP with the HEMT region protected by photoresist. Source/Drain ohmic contact of the HEMT was formed by e-beam evaporation of Ti/Al/Ni/Au and rapid annealing at 850°C for 30 s in N_2. The current spreading layer (CSL) of the LED was formed by e-beam evaporation of Ni/Au and RTA annealing at 570°C for 5 min in an air atmosphere [60]. Then, Ti/Al/Ti/Au metal layers were evaporated as p- and n-electrodes of the LED and patterned by lift-off. Subsequently, Ni/Au gate metallization of the HEMT was realized. The electrical connection between the HEMT and LED devices was implemented by metal wire-bonding.

The output curves ($I_d - V_d$) of the HEMT and LED are shown in Fig. 2.18(b), and the maximum output current and on-state resistance (R_{on}) are 9.2 mA (920 mA/mm) and 258 Ω (2.58 Ω/mm), respectively. The gate width (W_g) and gate length (L_g) are 10 and 1 μm, respectively. The forward voltage (V_F) of the 300×300 μm LED is 3.1 V under an injection current of 10 mA.

Although the performances of the HEMT and LED are comparable with the performances of the separately fabricated HEMT and LED, the potential limit of this strategy is the thin HEMT isolation buffer ($\sim$ 375 nm) between the channel and the conductive p-GaN (Fig. 2.16(b)), which will cause a large parasitic capacitance and low breakdown voltage for the HEMT on the top. As previously stated, increasing the buffer thickness may degrade the LED performance due to a large thermal budget. The capacitance-voltage (C-V) characteristics of the HEMTs grown on LED-epi and the HEMTs grown on bare sapphire were characterized. The maximum capacitance (C_{max}) of both HEMTs are approximately 2.77 nF, whereas the minimum capacitance (C_{min}) of the HEMT on the LED was approximately 100 pF, which is almost two orders of magnitude higher than the C_{min} (4.4 pF) of the HEMT on sapphire. Because the 2DEG shields the electric field at voltages higher than V_{th}, the capacitance of the HEMT structure dominated, and C_{max} remained constant. The increased C_{min} is caused by the relatively thin HEMT buffer layer (375 nm) and the highly conductive LED layer underneath. The C_{gb} and C_{bs} are the capacitances related to the p-GaN body layer and are significantly smaller than C_{gs} under a pinch-off condition (Fig. 2.19). Consequently, C_{min} is determined by C_{gb}, C_{bs}, and the buffer thickness. Because an additional leakage path exists between the gate and the drain via the p-GaN layer, a thicker buffer is also preferred from the breakdown voltage perspective. However, a substantially thicker buffer layer accompanied by a prolonged growth time will degrade the LED multi-quantum-well (MQW) quality and related V_F. This situation is a trade-off between the buffer thickness and V_F. As a result, a 375-nm buffer and a lower growth temperature are utilized in [58].

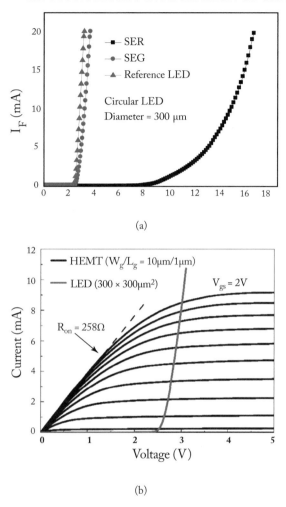

Figure 2.18: (a) I-V characteristics of LEDs fabricated by SER, SEG, and the reference LED; (b) output characteristics of the HEMT device and LED.

The off-state breakdown voltages (BV_{off}) of the HEMT-LED with various L_{gd} are below 40 V (Fig. 2.19(b)). The measured buffer breakdown voltage was 42.5 V. The HEMT-LED yields a lower breakdown voltage with a larger leakage current compared with typical HEMTs grown on sapphire substrates. The breakdown voltage was primarily limited for two reasons. The first reason is the parasitic leakage channel of the p-GaN layer. The second reason is the high background carrier density of the UID-GaN, which is caused by the lower growth temperature.

The control voltage V_{dd} and the gate modulation voltage were applied to the HEMT-LED with the circuit configuration shown in the inset diagram of Fig. 2.16(b). The HEMT-LED device emitted modulated blue light with a wavelength of 470 nm under a modulated gate bias. The forward current and light output power (LOP) versus the applied voltages are plotted in Fig. 2.20. The HEMTs (W_g/L_g =100/1 μm) exhibited reasonable control over the luminosity and injection current of the LED.

Although preliminary progress has been achieved for the SER of the HEMT on the top of the LED, some optimization is required to enhance the device breakdown voltage, although not at the cost of degraded LED performance. The alternative is to selectively grow an LED on the top of the HEMT. The n^+-GaN in the LED can be employed as the source/drain contacts in the HEMT. Additional improvement is anticipated in the future.

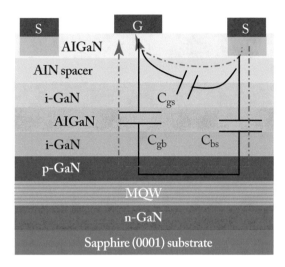

(a)

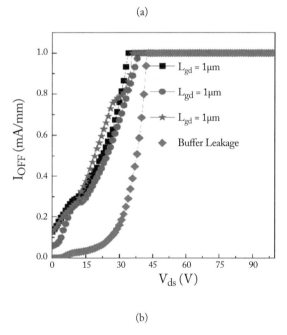

(b)

Figure 2.19: (a) Cross-sectional view of the circular Schottky gate diode of the HEMT-LED for the C-V measurements. The dotted red lines represent the potential leakage paths. C_{gs}, C_{gb}, and C_{bs} are the gate-source, gate-body, and body-source capacitances, respectively. The spacing between the gate and the source is approximately 15 μm. (b) BV_{off} of the HEMT-LED device with varied source-drain spacing. The inset diagram shows the measured buffer structure.

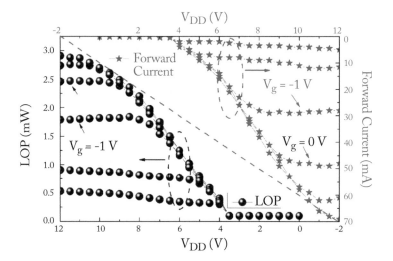

Figure 2.20: HEMT-LED device with modulated light emission under gate-bias and modulated I-V and LOP-V characteristics of the HEMT-LED device.

CHAPTER 3

III-V Materials and Devices

3.1 III-V HETEROSTRUCTURES FOR HIGH ELECTRON MOBILITY TRANSISTORS

The high-electron-mobility transistor represents one of the most important device applications for III-V heterostructures. The transistor is based on the heterojunction that is formed by a minimum of two semiconducting materials with different band gaps. Due to their relative band alignments, band discontinuities occur at the hetero-interface between the two semiconducting materials, which can be quantified using the conduction and valence band offsets ΔE_c and ΔE_v. As a result, a potential well is formed at the hetero-interface, which confines electrons in the horizontal direction. Within the potential well, the electrons can only move freely in the plane parallel to the hetero-interface; therefore, they are referred to as a two-dimensional electron gas (2DEG). A typical III-V HEMT structure is composed of an unintentionally doped (UID)-Schottky barrier, a carrier supplied layer, a UID-spacer and a UID-channel layer. The barrier, carrier supply layer, and spacer are formed by a wide bandgap semiconductor, whereas a narrow bandgap semiconductor with high intrinsic electron mobility usually serves as the channel. By separating the carrier supply layer and the electron transport channel, the ionized impurity scattering was dramatically reduced to simultaneously achieve high electron mobility and high carrier density.

The diverse III-V semiconductors offer an extensive variety of heterostructures that can be used to produce quantum wells for 2DEG confinement. Over the past few decades, high-electron-mobility transistor technology has been extensively developed using heterojunctions that are based on AlGaAs/GaAs [61], AlGaAs/InGaAs [62], InAlAs/InGaAs [63], AlSb/InAs [64], and AlSb/InSb [65]. Concurrent with the development of HEMTs, modulation-doped two-dimensional hole gases and high-hole-mobility transistors have also been realized in several heterostructures systems such as AlGaAs/GaAs [66], GaAs/InGaAs [67], InAlAs/InGaAs [68], AlAsAs/GaSb [69], and AlGaSb/InGaSb [70].

To produce high-performance HEMTs, parameters such as band offset, alloy composition, doping concentration and epi-layer thickness should be well designed. In an HEMT, the gate electrode modulates the conducting 2DEG channel through the Schottky contact formed between the gate metal and the barrier. The Schottky contact, which exhibits a rectifying characteristic, prevents large leakage current from flowing through the gate to the channel. Therefore, a semiconductor with a larger bandgap and a higher Schottky barrier height is preferred to suppress gate leakage. In addition, the thickness of the Schottky barrier/spacer can significantly impact the device performance. Because the physical gate length can currently be scaled down to the sub-100-

nm regime, a thin barrier between the III-V channel and the gate metal is essential for ensuring good electrostatic control and reducing the access resistance across the Schottky barrier at the source/drain region. However, the electron mobility of the 2DEG may be compromised [71].

Nominally, the channel layer in an HEMT structure should be lattice matched or almost lattice matched to the barrier/spacer layer. However, to enhance the electron transport properties, an ultra-thin mismatched channel layer, which possesses higher intrinsic electron mobility but a different lattice constant, can be employed. During device growth, the thickness of this strained channel layer is controlled within the critical thickness to prevent the generation of misfit dislocations at the hetero-interfaces. The HEMTs with this channel design are referred to as "pseudomorphic" HEMTs (pHEMTs) [62]. Generally, pHEMTs exhibit larger conduction band offsets and higher electron mobility compared with lattice-matched counterparts that utilize the same material system. The channel thickness is another important aspect. For devices with highly scaled gate lengths, a thin channel layer is preferred to increase the channel aspect ratio (gate length divided by the total thickness of the channel and the barrier layer) and reduce output conductance [72], which is desirable in terms of the RF characteristics.

In an HEMT, the electrons confined in the quantum well are separated from their parent donor atoms in adjacent barrier layers by modulation doping. Because no ionized impurity atoms exist to collide within the well, the electrons are free to move with high mobility along the quantum-well layer. The achievable 2DEG density is determined by the inherent properties of the heterostructures such as the conduction band discontinuity, the spacer thickness and the doping concentration [73]. High carrier densities in the quantum wells are desirable for improving the transconductance of the transistors. Modulation doping can be achieved using two approaches: homogeneous doping and δ-doping. The δ-doping technique represents the ultimate limit for scaling doping profiles. For nano-scale gate length devices, this scaling of doping profiles is important for obtaining better control of the sheet electron density in the channel. Compared with conventional homogeneous doping, δ-doping enables higher electron density in the channel and higher transconductance of the transistors [73]. In conventional HEMTs, the doping profile is introduced in the upper Schottky barrier. However, if the doping is placed within the backside wide bandgap barrier under the undoped channel, the resulting HEMT structure is referred to as an "inverted" HEMT [74]. For the inverted structure, the Schottky metal gate can be directly placed on the undoped channel layer to control the 2DEG formed at the reversed heterointerface. The concept of an inverted HEMT structure emerged almost simultaneously with the notion of a normal HEMT [61]. However, the inverted HEMT structure has received less attention and has not been investigated with the same intensity. Recently, an increased interest in combining the inverted HEMT structure with high-k dielectrics has developed toward realizing III-V high-mobility quantum well channel metal-oxide-semiconductor field effect transistors for high-speed, low-power logic applications [75, 76].

To evaluate the material quality and structural properties of HEMTs, figures of merits such as surface morphology, stress incorporation, defect density, sheet carrier density, and electron mobility are useful measures. Electron mobility is one of the most important criteria because it reflects the carrier transport properties in the quantum wells. The electron mobility is directly linked to the carrier scattering mechanisms. Note that the dominant carrier scattering mechanism in a bulk semiconductor can greatly differ from the dominant carrier scattering mechanism in an HEMT. This difference can be assessed by measuring the temperature dependency of the mobility. In a semiconductor material, the mobility is a measure of the responsiveness of carrier motion to an electric field. In a low electric field, the mobility can be correlated to the carrier drift velocity according to

$$V_{drift} = \mu E,$$

where μ is the mobility and E is the electrical field.

A simple model provides the approximate relation between scattering time (average time between scattering events) and mobility. After each scattering event, the carrier's motion is assumed to be randomized; thus, it has zero average velocity. Subsequently, the carrier uniformly accelerates in the electric field until it scatters again. The resulting average drift mobility is

$$\mu = \frac{q}{m^*}\tau,$$

where q is the elementary charge, m^* is the carrier effective mass and τ refers to the average scattering time.

In a bulk semiconductor, various scattering mechanisms limit the total mobility. Typical scattering mechanisms include ionized impurity scattering, neutral impurity scattering, acoustic phonon scattering and optical phonon scattering. The Coulombic interaction between a carrier and a charged atom causes ionized scattering. Therefore, ionized impurity scattering increases as the concentration of impurities increases. Neutral impurity scattering is weaker than ionized impurity scattering. As the thermal velocity of carriers increases with high temperature, ionized impurity scattering becomes less relevant at high temperature as the interaction time decreases. The temperature dependence for ionized-impurity-scattering-limited mobility (μ_{II}) can be expressed as follows:

$$\mu_{II} \propto T^{3/2}.$$

Phonon scattering is also referred to as lattice scatting. In a real crystal, atoms are not fixed at rigid sites on a lattice; they vibrate. A phonon is a "particle" that is associated with the thermal vibration of a lattice. When the adjacent two atoms on the unit cell vibrate along the same direction, this is referred to as acoustic vibration. When the adjacent two atoms on the unit cell vibrate in opposing directions, this is referred to as optical vibration. Four types of phonons exist: the transverse acoustical type, the longitudinal acoustical type, the transverse optical type and the longitudinal optical type. In the transverse mode, atoms vibrate in the direction that is perpendicular to a wave vector. In the optical mode, atoms vibrate along the direction of the wave vector. An optical phonon is more energetic than an acoustic phonon. Optical phonon scattering occurs when carriers possess substantial kinetic energy. Acoustic phonon scattering ($\propto 1/\mu_{AC}$) increases with temperature. More acoustic phonons are observed at high temperatures. The temperature dependence for acoustic-phonon-scattering-limited mobility is expressed as

$$\mu_{AC} \propto T^{-3/2}.$$

By applying various scattering mechanisms, the total carrier mobility in a semiconductor can be expressed by Matthiesen's rule:

$$\frac{1}{\mu} = \sum_i \frac{1}{\mu_i}.$$

In a bulk semiconductor, such as high purity GaAs [77], scattering by optical phonons dominates the mobility at temperatures > 100 K. Ionized impurity scattering limits mobility in the ultra-low temperature range. Scattering by neutral impurities is significantly weaker than scattering by the ionized impurity atoms due to the lack of Coulombic charge. Deformation potential and piezoelectric scattering are minor scattering mechanisms.

In a high-electron-mobility transistor, the electrons that are supplied by the donors in the barrier move into the potential well in the undoped channel. Specific features distinguish the electron transport in two-dimensional electron gas systems and bulk materials. Considering the ionized impurity scattering, the electrons in an HEMT can be scattered by remote impurities located within the doped region in the barrier and by residual background impurities in the channel. For HEMTs that utilize ternary compounds, additional scattering by alloy-disorder can be significant. W. Walukiewicz et al. analyzed the electron mobility in $In_{0.53}Ga_{0.47}As$-$Al_{0.52}In_{0.48}As$ with a 8-nm spacer width [78]. At high temperatures above 100 K, optical phonon scattering limits the total mobility. At low temperatures, alloy disorder scattering serves a crucial role and limits the two-dimensional electron gas mobility.

The high-electron-mobility transistor was first demonstrated at Fujitsu Labs in 1980 [61]. Over the course of more than 30 years of development, these high-speed, high-frequency transistors have established their niche in performance-driven applications and cannot be matched by standard Si CMOS technology. Figure 3.1 shows the record f_T for GaAs MESFETs, GaAs PHEMTs and InP HEMTs as a function of year of demonstration [79]. InAlAs/InGaAs HEMTs on InP have particularly excelled at extreme levels of noise and frequency and have yielded the optimal high-frequency performance for more than twenty consecutive years. The ever-increasing high-frequency response has been achieved through efforts devoted to device structure optimization and improvements in device processing technology, including scaling the gate length, increasing the InAs content of the InGaAs channel, reducing the barrier and/or channel thickness, reducing the parasitic resistances and capacitances and improving the short-channel effects. Using a laterally scaled-down tiered-edge ohmic structure, H. Matsuzaki demonstrated a 50-nm-gate-length $In_{0.53}Ga_{0.47}As$/InAs composite channel HEMT on InP, which exhibits extrinsic transconductance of 2.0 S/mm, extrinsic f_T of 496 GHz and ultra-low ohmic resistance of 100 $\Omega \cdot \mu m$ [80]. T.W. Kim et al. developed a self-aligned gate technology for InGaAs

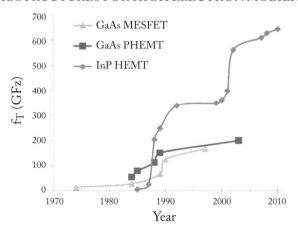

Figure 3.1: Short-circuit current-gain cut-off frequency (f_T) for GaAs MESFETs, GaAs PHEMTs and InP HEMTs as a function of year of demonstration [79].

high-electron-mobility transistors with non-alloyed Mo-based ohmic contacts and a very low parasitic capacitance gate design [81]. The new process delivered a contact resistance of 7 ohm·μm and a source resistance of 147 ohm·μm. R. Lai et al. [82] was the first researchers to demonstrate a sub-50-nm InP HEMT device with f_{max} greater than 1 THz. D.H. Kim et al. [83] reported a 50-nm E-mode $In_{0.7}Ga_{0.3}As$ PHEMTs on a 100-mm InP substrate with $f_{max} > 1$ THz by combining a Pt gate sinking process and a two-step recess process to effectively scale-down the $In_{0.52}Al_{0.48}As$ barrier layer. A. Leuther et al. [84] experimentally demonstrated metamorphic HEMTs (mHEMTs) with a gate length of 20 nm on GaAs substrates with a strained $In_{0.8}Ga_{0.2}As$ channel. The realized mHEMTs achieved a source resistance R_S of 100 Ω·μm in addition to an extrinsic maximum transconductance of 2500 mS/mm and a maximum cut-off frequency f_T of 660 GHz. By employing a pure InAs core layer sandwiched between $In_{0.53}Ga_{0.47}As$ cladding layers, D.H. Kim and J. A. del Alamo successfully produced 30-nm InAs PHEMTs characterized by $f_T = 644$ GHz and $f_{max} = 681$ GHz [85]. These are the first transistors of any type of material system to simultaneously exhibit f_T and f_{max} values in excess of 640 GHz. A particularly remarkable finding is that this result was obtained at an operating voltage of 0.5 V, which showcases the outstanding low-power potential of this technology [79]. Using a 5-nm InAs thin channel, E.-Y. Chang recently enhanced the cut-off frequency f_T to 710 GHz for 60-nm-gate-length InAs HEMTs [72]. The InGaAs HEMT technology is uniquely suited for mm- and sub-mm-wavelength applications in radar, radio astronomy, high-resolution imaging arrays, high-capacity wireless communications, high-data-rate optical fiber systems and sub-mm-wavelength spectroscopy. InP-based high-speed, high-frequency transistors will continue to serve an important role in emerging terahertz applications.

3.2   III-V HIGH-MOBILITY CHANNEL MOSFETS

To date, Si CMOS technology has enabled state-of-the-art microprocessors with large-volume, low-cost production and high-level integration. The remarkable advances of the integrated circuits have been driven by the downscaling of Si MOSFETs following "Moore's Law," which states that the transistor count in a CPU increases by 2× and transistor feature size decreases by 0.7× approximately every two years. With the scaling of Si CMOS devices moving into the sub-22-nm regime, significant short-channel effects have posed many challenges. Increased device densities and circuit functionality have resulted in increased power consumption per chip. The power density dissipated by modern logic chips has already hit the limitation at approximately 100 W/cm² [86], and CMOS technology has entered a "power-constrained" scaling era. Under these circumstances, the International Technology Roadmap for Semiconductors (ITRS) anticipates that the transistor supply voltage will be significantly reduced in future technology nodes due to its ability to quadratically and linearly reduce transistor active power and standby power, respectively. However, a reduction in the transistor operating voltage would compromise the speed of traditional Si transistors because the drive current would be significantly decreased at a reduced supply voltage. To resolve the competing constraints of high-speed and low-power integrated circuits, CMOS technology may experience by far the most radical reform: replacing the channel with a non-Si material in which carriers can travel at a much higher velocity. It has generated significant interest and research investment in exploring III-V high mobility channel transistors for logic applications. As shown in Table 3.1, III-V semiconductors generally exhibit significantly higher intrinsic electron mobility and smaller effective electron mass relative to Si or Ge. Currently, InGaAs, with a maximum electron mobility up to 10,000 cm²/V•s, has emerged

	Si	Ge	GaAs	InP	$In_{0.53}Ga_{0.47}As$	InAs	InSb
E_g (eV)	1.1	0.66	1.4	1.34	0.75	0.35	0.17
μ_n (cm²/V·s)	1350	3900	8000	4600	12000	40000	77000
μ_p (cm²/V·s)	480	1900	400	150	350	<500	850
m^*/m_0	0.165	0.12	0.067	0.082	0.041	0.024	0.014

Table 3.1: Basic electronic properties, including bandgap (E_g), electron mobility (μ_n), hole mobility (μ_p), and effective electron mass (m*) relative to the free electron mass of typical III-V semiconductors compared with Si and Ge

as the most promising new channel material to replace Si in nMOSFETs. Although other III-V alloys, such as InSb, exhibit even higher electron mobility, their small bandgaps may cause high band-to-band tunneling leakage and limit the transistor operating voltage in practical applications. In addition, InGaAs-based transistor technology has been the focus of studies and has held the record for high-frequency performance for more than twenty consecutive years. The sufficiently high mobility and mature processing technology render it the most competitive choice for high-speed, low-power logic circuits.

Conventional InP-based high-speed device technology has been well developed. However, to replace the Si channel with high mobility InGaAs for future high-speed, low-power, advanced CMOS, novel device technologies are required, and the important aspects of logic applications must be understood. In contrast to the functions of an amplifier in traditional analog, microwave or millimeter-wave applications, a transistor in logic circuits operates as a switch.

Figure 3.2 illustrates the most important figures of merit of a transistor that operates as a switch [87]. A state-of-the-art transistor in logic circuits should have a maximized drive current

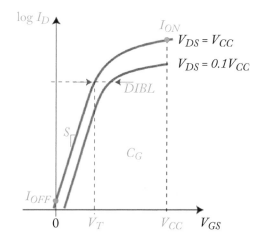

Figure 3.2: Electrical figures of merit of a transistor as a switch [87].

(I_{on}), a minimized off-state current (I_{off}), a steep transition between the on-state and the off-state, a low operating voltage (V_{cc}), a small variation in the threshold voltage (V_{th}) for different drain biases (V_{ds}), and a deeply down scaled device footprint. To achieve a viable III-V logic technology, a number of critical technological challenges need to be overcome. The following section will briefly review the recent progress that has been achieved to address these challenges.

In a traditional III-V high-electron-mobility transistor, a gate metal is directly placed on a wide bandgap barrier to form a Schottky contact. Although nanometer-scale, Indium-rich $In_xGa_{1-x}As$ or InAs channel HEMTs have demonstrated excellent logic figures of merits [88, 89], the large gate leakage and huge device footprint are not suitable for CMOS logic applications. A proper gate dielectric with a low interface state density and reasonable thermal stability is required. These characteristics were extremely difficult to achieve for III-V semiconductors due to the Fermi-pinning issue at the oxide/III-V interface until the recent breakthrough in atomic-layer-deposited (ALD) high-k dielectrics [90]. Specifically, the main challenges concerning the gate stack of InGaAs-based MOSFETs are as follows: (1) scaling the equivalent oxide thickness (EOT) down to 1 nm with acceptable gate leakage current to enable good gate electrostatic control; (2) reducing oxide/III-V interface state density below 1×10^{12} cm^{-2}· eV^{-1} to enable effective gate modulation and rapid on/off switching; and (3) realizing a sharp and flat interface and high effective channel mobility to provide high-current and high-speed performance. Substantial effort has been directed at developing advanced gate dielectric technologies, including *in situ* molecular beam epitaxy growth of Ga_2O_3, Gd_2O_3, and $Ga_2O_3/(Gd_xGa_{1-x})_2O_3$ [91–93], atomic layer deposition of Al_2O_3 [94], HfO_2 [95], ZrO_2 [96], and $ZrO_2/LaAlO_x$ [97], jet vapor deposition of Si_3N_4 [98], wet thermal oxidation of InAlP [99] and composite high-k gate stack of $TaSiO_x/InP$ [100]. Various surface passivation/pretreatment methods, including a Si interfacial passivation layer [101], an aluminum oxynitride interfacial passivation layer [102], plasma PH_3 passivation [103], phosphorous nitride interface passivation [104], silane-ammonia surface passivation [105], HBr pretreatment [106], and $(NH_4)_2S$ passivation [107], have also been actively investigated.

For the development of InGaAs channel nMOSFETs as a part of future logic technology, a proper channel design with high electron mobility and good compatibility with the CMOS fabrication process is essential. With the progress achieved on the gate dielectric for III-V materials, various device architectures are being actively explored, and significant progress has been obtained. A promising candidate is inversion-mode architecture [94, 106, 108]. The inversion mode architecture is a III-V embodiment of a traditional bulk Si MOSFET. The Si channel is replaced by high-mobility III-V materials such as $In_{0.53}Ga_{0.47}As$. An inversion channel is formed at the interface between the high-mobility III-V layer and the gate dielectric, whereas the source and drain regions are formed by ion implantation. The advantages of this architecture are that it is well known to and understood by the mainstream Si manufacturing community and that it should exhibit a scalability that is similar to bulk Si MOSFETs. However, because the channel is formed at the dielectric and semiconductor interface, it usually suffers from interface roughness and Coulomb scattering, as well as remote phonon scattering from high-k oxide, which severely degrade the effective channel mobility [109]. In contrast to the inversion mode MOSFETs, an alternative structure is the InGaAs "MOSHEMT-like" transistor [76, 100], which features buried quantum-well channels with high-k dielectric. A wide bandgap semiconductor barrier is usually inserted between the oxide and the channel. The increased proximity between the channel and the gate dielectrics and the heterojunction structure are beneficial for high electron mobility. The superior mobility in a buried channel devices over their surface channel counterparts was evidenced by a number of experimental results. In addition, the heterostructures in the buried channel design offer flexibility in band structure engineering and biaxial strain incorporation.

Source/drain engineering also serves an important role in nano-scale transistors. In an ultra-scaled technology node, the parasitic resistance caused by the source/drain junction and metal ohmic contacts can significantly degrade the device performance [110]. Previous studies have revealed that a highly doped source/drain with a maximum concentration of 10^{20} cm^{-3} is required to prevent "source starvation" for III-V MOSFETs to outperform Si and Ge MOSFETs under quasi-ballistic transport [111]. To enable the 11 nm technology, the ITRS targets a parasitic resistance R_{ext} of 110 $\Omega \cdot \mu$m for fully depleted high-performance devices [112]. In a traditional III-V high-electron-mobility transistor, a highly doped source/drain contact layer is usually deposited during material epitaxy. However, a recessed gate technique has to be employed during device fabrication to remove this contact layer under the gate region, which is incompatible with VLSI fabrication due to the resulting large footprint and high parasitic capacitances. In addition, the gate recess process gives rise to concerns about threshold voltage uniformity (caused by variations in recess etching depth) and channel damage beneath the recessed region. On the other hand, the physical properties of III-V materials are such that the source/drain implanted junction and silicide process, which have adequately served Si CMOS production, are not suitable for these compound semiconductors. Although inversion-mode InGaAs MOSFETs with implanted source/drain have shown various promising characteristics, these devices generally experience large access resistances. The large on-state resistances in these implanted devices are attributed to the following factors. First, unlike Si, the III-Vs alloys have a much lower thermal budget, which limits the post-implantation annealing temperature. Consequently, the source/drain dopants cannot be efficiently activated, and the maximum source/drain doping concentration is 1×10^{19}/cm^3 at best. Second, the crystalline damage caused by the implantation cannot be completely annealed by the post-implantation annealing process. To address these problems, alternative techniques have been pursued by various groups. Selective regrowth is an attractive technique and has been intensively studied. One major advantage to this technique is its ability to provide a heavily doped source/drain without significant crystal damage or any post-annealing as needed for ion implantation. A high doping concentration of 1×10^{20}/cm^3 is achievable. Moreover, in traditional III-V transistors, biaxial-strain is usually applied to increase performance. By introducing uni-axial stress in InGaAs MOSFET using lattice-mismatched source/drain regrowth, electron mobility and drive current can be improved. Therefore, selective regrowth is also a promising approach for strain engineering in device design. M. J. W. Rodwell et al. developed selective source/drain regrowth technique in III-V high mobility MOSFETs [113] and successfully applied it to the fabrication of nano-scale InGaAs and InAs surface channel MOSFETs [5, 110, 114, 115]. M. Egard et al. reported a high-transconductance, self-aligned gate-last surface channel In$_{0.53}$Ga$_{0.47}$As MOSFET using the MOCVD regrowth of highly doped source and drain access regions [75]. A 55-nm-gate-length In$_{0.53}$Ga$_{0.47}$As MOSFET exhibited an extrinsic transconductance of 1.9 mS/μm and on-resistance of 199 $\Omega \cdot \mu$m. K. M. Lau et al. developed high-performance InGaAs buried quantum well MOSFETs that were metamorphically grown on GaAs and Si substrates with MOCVD source/drain regrowth [76, 116–118]. A high drive current, high extrinsic transconductance and significantly low on-resistance of 129 $\Omega \cdot \mu$m were achieved. In parallel with the development of the source/drain regrowth technique, S. Takagi et al. developed a Ni-InGaAs metal S/D scheme for realizing scaled future III-V MOSFETs [119, 120]. A low source/drain parasitic resistance value of 274 $\Omega \cdot \mu$m in InAs-

OI nMOSFETs and 20-nm-L_{ch}, high-performance ETB InAs-OI MOSFETs on Si have been demonstrated with Ni-InGaAs metal S/D [120].

In 2011, Intel announced that their 3-D Tri-gate transistors (or "FinFETs") were advancing into high-volume production at the 22-nm node, which was marked as a significant breakthrough in the evolution of the transistor. In a FinFET, the gate electrode wraps around three sides of a vertical thinfin structure to form parallel conducting channels. The enhanced gate electrostatic control results in strong immunity to short-channel effects and significantly reduces power consumption. III-V high-mobility-channel MOSFETs may also be combined with the advanced 3D multi-gate structure if the planar design fails to satisfy the requirements. Some notable advances in this area should also be noted. Radosavljevic et al. [121] demonstrated 3D tri-gate InGaAs quantum well field-effect transistors with high-k dielectric and scaled gate-to-source/gate-to-drain separation. For the first time, the electrostatics of 3D InGaAs devices outperformed ultra-thin body planar InGaAs MOSFETs. Gu et al. [122] fabricated the first inversion-mode gate-all-around InGaAs MOSFETs on InP substrates. Good stability down to a channel length of 50 nm was demonstrated, owing to the good electrostatic control of the gate-all-around structure. Xiao et al. [123] have reported a controlled growth method for self-aligned planar GaAs nanowires via Au-catalyzed vapor-liquid-solid mechanism. They applied this technique to design and fabricate AlGaAs-GaAs nanowire HEMTs, which demonstrated excellent, uniform and scalable dc characteristics. Tomioka et al. [124, 125] have demonstrated position-controlled growth of vertical InGaAs nanowires on Si (111) substrates without any buffering technique. InGaAs nanowires and InGaAs/InP/InAlAs/InGaAs core-multishell nanowires were grown on SiO_2 patterned Si (111) substrates by selective area growth using MOCVD. Surrounding-gate transistors using core-multishell nanowires as channels exhibited enhanced on-state current and transconductance while maintaining good gate controllability. This configuration provides a route to the fabrication of vertically oriented transistors.

3.3   MONOLITHIC INTEGRATION OF III-V FETS ON SI

After half of a century of explosive development following Moore's Law, the scaling of Si-based ICs has reached nanometer physical gate lengths. It is generally recognized that further reduction is running into many technological challenges. Meanwhile, III-V compound semiconductors have been under development since the 1960s for high-frequency and photonic applications that simply cannot be matched by Si electronics. Significant progress in the development of InGaAs-based nMOSFETs for future high-speed, low-power logic applications has been recently achieved. However, Si has been and will remain the workhorse in the semiconductor industry. Looking forwards, it becomes increasingly appealing to monolithically integrate the III-V high-speed electronics with standard CMOS on a common Si platform. First, such integration is beneficial from the perspective of substrate engineering because the Si substrates are available with large diameters, lower cost and superior thermal conductivity and mechanical properties. Second, combining the advantages of the high-level integration and high-volume production of Si-based electronic circuitry with the superior high-speed performance of III-V components can pave the way toward a new class of hybrid integrated circuits. By placing III-V RF, communication and logic blocks adjacent to traditional Si CMOS cells and connecting the devices and subcircuits with standard on-wafer interconnects, a system-on-chip platform can be realized. In this section, the major challenges of and recent progress on the heterointegration of III-V electronic materials and devices on Si will be reviewed.

To enable high-performance III-V electronics on a large-diameter Si platform, achieving high-quality III-V thin films on Si substrates is essential. However, the integration of these dissimilar semiconductors on Si has been challenging. Single crystal (001) Si exhibits a diamond structure, whereas (001)-oriented III-V semiconductors generally crystalize in a zinc blende structure. The reduced crystal symmetry in III-V materials causes atom stacking disorders and planar defects in material growth. Moreover, the lattice mismatch between III-V semiconductors and Si causes various types of defects in the heteroepitaxial layers. In addition, the thermal expansion coefficient and the thermal conductivity between III-V semiconductors and Si differ. The thermal expansion coefficients of GaAs and InP are 2.6× higher and 1.8× higher, respectively, than the thermal expansion coefficients of Si.

To obtain III-V thin films on Si, wafer bonding and heteroepitaxy are the two major research directions that are being actively explored. The wafer bonding technique is able to transfer highquality III-V ultra-thin films to a Si substrate covered by a dielectric regardless of any lattice mismatch. Rapid progress is being achieved in this field. Yokoyama et al. [126] reported the formation of ultra-thin-body III-V-semiconductor-on-insulator (III-V-OI) nMOSFETs on Si by low-damage direct wafer bonding (DWB), in which the InGaAs channel thickness was successfully reduced to 7 nm. Using the DWB technique, integration of high-mobility InGaAs n-channel and Ge p-channel MOSFETs with self-aligned Ni—InGaAs and Ni—Ge metal source/drain on a Ge substrate was also demonstrated [127]. Another research group developed an epitaxial transfer method for the integration of an ultra-thin single-crystal InAs layer on a Si/SiO$_2$ substrate [128]. Top-gated InAs channel FETs with well-behaved device characteristics were demonstrated using compound semiconductor-on-insulator technology. III-V CMOS circuits that consist of an InAs channel nMOS and an InGaSb channel pMOS on Si were realized [129]. However, all the layer-transfer techniques suffer from the difficulties in achieving adequate uniformity and high yield and are limited by the size of the III-V host wafers.

In contrast to wafer bonding, heteroepitaxy can take advantage of large wafer size, good thermal conductivity and mechanical robustness of Si substrates, suitable for low-cost mass production. The main concerns of this technique are the defect density, buffer thickness and thermal conductivity. The high-density defects are caused by the lattice mismatch, thermal expansion coefficient mismatch, and the growth of polar semiconductors on non-polar substrates. The lattice mismatch generates misfit dislocations at the heterointerface and introduces compressive stress in the III-V thin films. The propagation of misfit dislocations away from the heterointerface (the threading dislocations) can significantly degrade device performance. During heteroepitaxy, the thermal mismatch causes additional stress in the temperature ramp-up and reactor cool-down processes. In addition to the previously mentioned threading dislocations, other planar defects, such as anti-phase domains, twin defects and stacking faults, may also cause large leakage currents and device failure. Successful attempts at growing InP related heterostructures of device quality on planar Si substrates by MOCVD or MBE have been achieved. Traditional blanket heteroepitaxy on planar substrates usually involves thick composite buffers, such as AlSb/GaSb/GaP [130], InP/GaAs [131], InAlAs/GaAs/Ge [132] and InAlAs/GaAs [133], for dislocation management. Figures 3.3(a)–(d) present cross-sectional TEM images for InAs HEMT on Si using a GaSb/GaP accommodation layer [130], InGaAs metamorphic HEMT on Si using a InP/GaAs buffer [131], $In_{0.52}Al_{0.48}As/In_{0.7}Ga_{0.3}As$ quantum wells on Si using a composition-graded InAlAs/GaAs buffer [133] and InP-based HBTs on Ge-on-insulator-on-Si [132]. Figure 3.1(e) shows a top-view SEM of the InP HBTs integrated with Si CMOS on a common substrate with micron-length separations [132]. Currently, $In_{0.52}Al_{0.48}As/In_xGa_{1-x}As$ ($x \geq 53\%$) quantum wells with Hall mobility comparable to the quantum wells grown on lattice-matched InP substrates have been reported [134–136]. InAs HEMTs on Si substrates with room temperature Hall mobility above 27000 $cm^2V^{-1}s^{-1}$ [130, 137] have also been achieved. A drawback of the heteroepitaxy of III-V materials on blanket Si substrates is that fairly thick III-V buffers must be deposited on the entire Si substrate to satisfy the minimum defect density requirement, and the unnecessary areas are etched away for transistor fabrication. This process can be material-intensive, and the required long growth time limits the process throughput.

Compared with conventional heteroepitaxy on planar Si substrates, a unique alternative approach is nanopatterned growth. The advantage of nanopatterned growth is that low-defect-density films can be achieved with thin buffer layers. A FinFET transistor structure can be defined by the trench pattern for CMOS integration in a bottom-up fashion. The use of nanopatterned growth for the monolithic integration of Ge and III-V semiconductors on Si has been rapidly evolving in recent years. In 2000, Langdo et al. [138] reported selective epitaxial growth of high-quality Ge on SiO_2/Si substrates in 100-nm holes using ultra-high-vacuum chemical vapor deposition. Threading dislocations were blocked at the oxide sidewalls due to the "epitaxial necking effect." Park et al. [139] subsequently noted defect reduction in selective Ge epitaxy in growth windows formed by trenches on Si (001) substrates. Dislocations originating from the Ge/Si interface were completely trapped in trenches with maximum widths of 400 nm. Additional studies of the heteroepitaxy of GaAs and InP on patterned Si substrates have been performed. Li et al. [140] reported the selective epitaxial growth of GaAs on Si using submicron SiO_2 strip patterns with aspect ratios >1. The majority of the dislocations were guided to the SiO_2 sidewalls, and a defect-free top surface was obtained. The same group subsequently reported high-quality, uncoalesced InP selectively grown in sub-micron SiO_2 trenches on a Si (001) substrate using GaAs as an intermediate buffer [141]. A v-grooved Si surface was used to effectively trap misfit defects and anti-phase domains. Merckling et al. examined the selective area epitaxy of InP grown on 300-mm on-axis Si (001) substrates patterned with shallow trench isolation using rounded-

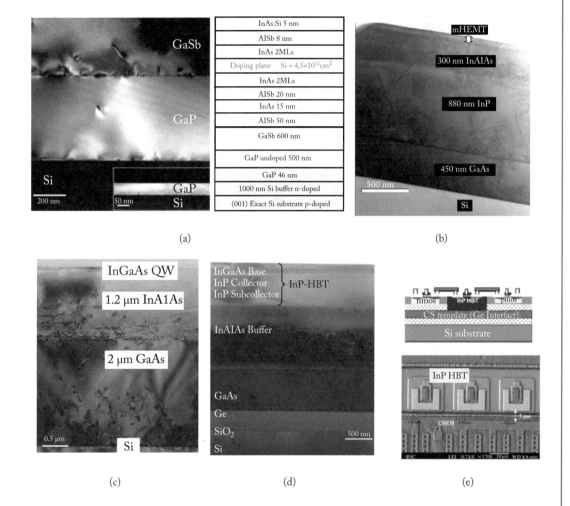

Figure 3.3: (a)–(d): TEM images of InAs HEMT on Si using a GaSb/GaP accommodation layer [130], InGaAs metamorphic HEMT on Si using InP/GaAs buffer [131], In$_{0.52}$Al$_{0.48}$As/In$_{0.7}$Ga$_{0.3}$As quantum wells on Si using a composition-graded InAlAs/GaAs buffer [133] and InP-based HBTs on Ge-on-insulator-on-Si [132]. (e) the top view SEM of the InP HBTs integrated with Si CMOS on a common substrate with micron-length separations [132].

Ge [142] or V-grooved Si surfaces [143, 144]. Cipro et al. [145] investigated the selective growth of GaAs/AlAs/InGaAs/AlAs/GaAs/Si multilayers using the aspect ratio trapping technique in SiO$_2$ cavities on 300-mm substrates. The InGaAs layers obtained with no anti-phase boundaries are perfect candidates for integration as channels in n-type metal oxide semiconductor field-effect transistor, and the low temperatures can allow for the co-integration with p-type MOSFETs.

CHAPTER 4

Summary

For GaN MIS-HEMT, the device technology that has used *ex situ* Al_2O_3 or *in situ* SiN_x to simultaneously suppress gate leakage current and neutralize surface traps is presented and analyzed. To explore the potential of GaN HEMTs in RF power applications, device configuration elements are correlated with several key device performance parameters. Four device scaling technologies, namely, source/strain regrowth, (In)AlN barrier thinning, T-gate fabrication, and gate-last self-aligned technology, are comprehensively reviewed. As an example of GaN device integration, preliminary progress on the integration of GaN HEMTs with LEDs, which paved a path for achieving future all-III-N solutions, is discussed.

Comprehensive coverage of III-V HEMT types and the status of the state of the art of III-V HEMTs are offered to demonstrate their unique advantages for millimeter-wave applications. An in-depth discussion on the possibility of replacing a Si channel with a III-V channel with an emphasis on several critical elements, including gate dielectrics, S/D engineering, and advanced gate patterns, is provided. The recent progress in heteroepitaxial growth on III-V materials and heterojunctions on Si, with the ultimate goal of integrating compound semiconductors with a mature Si platform, is also presented.

References

[1] V. Tilak, B. Green, H. Kim, R. Dimitrov, J. Smart, W. J. Schaff, J. R. Shealy, and L. F. Eastman, "Effect of passivation on AlGaN/GaN HEMT device performance," in *IEEE International Symposium on Compound Semiconductors*, Monterey, CA, pp. 357–363, 2000. DOI: 10.1109/ISCS.2000.947182. 3

[2] O. Jardel, F. De Groote, C. Charbonniaud, T. Reveyrand, J.-P. Teyssier, R. Quere, and D. Floriot, "A drain-lag model for AlGaN/GaN power HEMTs," in *IEEE/MTT-S International Microwave Symposium*, Honolulu, HI, pp. 601–604, 2007. DOI: 10.1109/MWSYM.2007.379972. 3

[3] H. Kim, R. M. Thompson, V. Tilak, T. R. Prunty, J. R. Shealy, and L. F. Eastman, "Effects of SiN passivation and high-electric field on AlGaN-GaN HFET degradation," *IEEE Electron Device Lett.*, vol. 24, pp. 421–423, 2003. DOI: 10.1109/LED.2003.813375. 3

[4] X. Lu, J. Ma, H. Jiang, and K. M. Lau, "Characterization of in situ SiNx thin film grown on AlN/GaN heterostructure by metal-organic chemical vapor deposition," *Appl. Phys. Lett.*, vol. 104, p. 032903, 2014. DOI: 10.1063/1.4862664. 3, 7, 8, 10

[5] L. Dong Seup, O. Laboutin, C. Yu, W. Johnson, E. Beam, A. Ketterson, M. Schuette, P. Saunier, and T. Palacios, "Impact of Al2O3 passivation thickness in highly scaled GaN HEMTs," *IEEE Electron Device Lett.*, vol. 33, pp. 976–978, 2012. DOI: 10.1109/LED.2012.2194691. 3, 39

[6] T. Huang, K. M. Wong, M. Li, X. Zhu, and K. M. Lau, "Effect of post-gate RTA on leakage current (Ioff) in GaN MOSHEMTs," *Phys. Status Solidi C*, vol. 9, pp. 919–922, 2012. DOI: 10.1002/pssc.201100444. 3, 4

[7] Z. Liu, G. Ng, H. Zhou, S. Arulkumaran, and Y. Maung, "Reduced surface leakage current and trapping effects in AlGaN/GaN high electron mobility transistors on silicon with SiN/Al2O3 passivation," *Appl. Phys. Lett.*, vol. 98, p. 113506, 2011. DOI: 10.1063/1.3567927. 3

[8] M. Anand, G. Ng, S. Vicknesh, S. Arulkumaran, and K. Ranjan, "Reduction of current collapse in AlGaN/GaN MISHEMT with bilayer SiN/Al2O3 dielectric gate stack," *Phys. Status Solidi C*, vol. 10, pp. 1421–1425, 2013. DOI: 10.1002/pssc.201300219. 3, 10

[9] C. Sanabria, A. Chakraborty, H. Xu, M. J. Rodwell, U. K. Mishra, and R. A. York, "The effect of gate leakage on the noise figure of AlGaN/GaN HEMTs," *IEEE Electron Device Lett.*, vol. 27, pp. 19–21, 2006. DOI: 10.1109/LED.2005.860889. 3

[10] A. Balandin, S. Morozov, S. Cai, R. Li, K. Wang, G. Wijeratne, and C. Viswanathan, "Low flicker-noise GaN/AlGaN heterostructure field-effect transistors for microwave communications," *IEEE Trans. Microwave Theory Tech.*, vol. 47, pp. 1413–1417, 1999. DOI: 10.1109/22.780388. 3

[11] P. Ye, B. Yang, K. Ng, J. Bude, G. Wilk, S. Halder, and J. Hwang, "GaN metal-oxide-semiconductor high-electron-mobility-transistor with atomic layer deposited Al2O3 as gate dielectric," *Appl. Phys. Lett.*, vol. 86, pp. 063501–063501-3, 2005. DOI: 10.1063/1.1861122. 4

[12] T. Huang, X. Zhu, K. M. Wong, and K. M. Lau, "Low-leakage-current AlN/GaN MOSHFETs using for increased 2DEG," *IEEE Electron Device Lett.*, vol. 33, pp. 212–214, 2012. DOI: 10.1109/LED.2011.2176909. 5, 6

[13] H. Wang, J. W. Chung, X. Gao, S. Guo, and T. Palacios, "Al2O3 passivated InAlN/GaN HEMTs on SiC substrate with record current density and transconductance," *Phys. Status Solidi C*, vol. 7, pp. 2440–2444, 2010. DOI: 10.1002/pssc.200983899. 5

[14] K. Čičo, J. Kuzmík, J. Liday, K. Hušeková, G. Pozzovivo, J.-F. Carlin, N. Grandjean, D. Pogany, P. Vogrinčič, and K. Fröhlich, "InAlN/GaN metal-oxide-semiconductor high electron mobility transistor with Al2O3 insulating films grown by metal organic chemical vapor deposition using Ar and NH3 carrier gases," *J. Vac. Sci. Technol., B*, vol. 27, pp. 218–222, 2009. DOI: 10.1116/1.3021034. 5

[15] K. Shinohara, D. C. Regan, T. Yan, A. L. Corrion, D. F. Brown, J. C. Wong, J. F. Robinson, H. H. Fung, A. Schmitz, T. C. Oh, S. J. Kim, P. S. Chen, R. G. Nagele, A. D. Margomenos, and M. Micovic, "Scaling of GaN HEMTs and Schottky diodes for submillimeter-wave MMIC applications," *IEEE Trans. Electron Devices*, vol. 60, pp. 2982–2996, 2013. DOI: 10.1109/TED.2013.2268160. 6, 13, 14, 15, 17

[16] S. C. Binari, K. Doverspike, G. Kelner, H. Dietrich, and A. Wickenden, "GaN FETs for microwave and high-temperature applications," *Solid State Electron.*, vol. 41, pp. 177–180, 1997. DOI: 10.1016/S0038-1101(96)00161-X. 6

[17] E. Alekseev, A. Eisenbach, and D. Pavlidis, "Low interface state density AlN/GaN MIS-FETs," *Electron. Lett.*, vol. 35, pp. 2145–2146, 1999. DOI: 10.1049/el:19991407.

[18] H. Kawai, M. Hara, F. Nakamura, and S. Imanaga, "AlN/GaN insulated gate heterostructure FET with regrown n+ GaN ohmic contact," *Electron. Lett.*, vol. 34, pp. 592–593, 1998. DOI: 10.1049/el:19980464. 6

[19] F. Medjdoub, M. Zegaoui, D. Ducatteau, N. Rolland, and P. Rolland, "High-performance low-leakage-current AlN/GaN HEMTs grown on silicon substrate," *IEEE Electron Device Lett.*, vol. 32, pp. 874–876, 2011. DOI: 10.1109/LED.2011.2138674. 6

[20] F. Medjdoub, J. Derluyn, K. Cheng, M. Leys, S. Degroote, D. Marcon, D. Visalli, M. Van Hove, M. Germain, and G. Borghs, "Low on-resistance high-breakdown normally off AlN/GaN/AlGaN DHFET on Si substrate," *IEEE Electron Device Lett.*, vol. 31, pp. 111–113, 2010. DOI: 10.1109/LED.2009.2037719.

[21] F. Medjdoub, M. Zegaoui, N. Rolland, and P. Rolland, "Demonstration of low leakage current and high polarization in ultrathin AlN/GaN high electron mobility transistors grown on silicon substrate," *Appl. Phys. Lett.*, vol. 98, p. 223502, 2011. DOI: 10.1063/1.3595943. 6

[22] T. Huang, X. Zhu, and K. M. Lau, "Enhancement-mode AlN/GaN MOSHFETs on Si substrate with regrown source/drain by MOCVD," *IEEE Electron Device Lett.*, vol. 33, pp. 1123–1125, 2012. DOI: 10.1109/LED.2012.2198911. 6

[23] M. Van Hove, X. Kang, S. Stoffels, D. Wellekens, N. Ronchi, R. Venegas, K. Geens, and S. Decoutere, "Fabrication and performance of Au-free AlGaN/GaN-on-silicon power devices with Al2O3 and Si3N4/Al2O3 gate dielectrics," 2013. DOI: 10.1109/TED.2013.2274730. 7, 12

[24] E. Cho, S. Seo, C. Jin, D. Pavlidis, G. Fu, J. Tuerck, and W. Jaegermann, "Impact of in situ SiNx layer grown with metal organic vapor phase epitaxy on the electrical and optical properties of AlN/GaN metal insulator semiconductor field effect transistor structures," *J. Vac. Sci. Technol., B*, vol. 27, pp. 2079–2083, 2009. DOI: 10.1116/1.3186615.

[25] X. Lu, J. Ma, Z. Liu, H. Jiang, T. Huang, and K. M. Lau, "In situ SiNx gate dielectric by MOCVD for low-leakage-current ultra-thin-barrier AlN/GaN MISHEMTs on Si," *Phys. Status Solidi A*, vol. 211, pp. 775–778, 2014. DOI: 10.1002/pssa.201300495. 7, 8, 12

[26] E. Ogawa, T. Hashizume, S. Nakazawa, T. Ueda, and T. Tanaka, "Chemical and potential bending characteristics of SiNx/AlGaN interfaces prepared by in situ metal-organic chemical vapor deposition," *Jpn. J. Appl. Phys.*, vol. 46, p. L590, 2007. DOI: 10.1143/JJAP.46.L590. 8

[27] S. Huang, S. Yang, J. Roberts, and K. J. Chen, "Threshold voltage instability in Al2O3/GaN/AlGaN/GaN metal–insulator–semiconductor high-electron mobility transistors," *Jpn. J. Appl. Phys.*, vol. 50, p. 0202, 2011. DOI: 10.7567/JJAP.50.110202. 10

[28] C. Mizue, Y. Hori, M. Miczek, and T. Hashizume, "Capacitance–voltage characteristics of Al2O3/AlGaN/GaN structures and state density distribution at Al2O3/AlGaN interface," *Jpn. J. Appl. Phys.*, vol. 50, p. 1001, 2011. DOI: 10.7567/JJAP.50.021001. 10

[29] D. G. Park, M. Tao, D. Li, A. E. Botchkarev, Z. Fan, Z. Wang, S. N. Mohammad, A. Rockett, J. R. Abelson, H. Morkoç, A. R. Heyd, and S. A. Alterovitz, "Gate quality Si3N4 prepared by low temperature remote plasma enhanced chemical vapor deposition for III–V semiconductor-based metal–insulator–semiconductor devices," *J. Vac. Sci. Technol., B*, vol. 14, pp. 2674–2683, 1996. DOI: 10.1116/1.589003. 10

[30] P. Kordos, R. Stoklas, D. Gregusova, and J. Novak, "Characterization of Al-GaN/GaN metal-oxide-semiconductor field-effect transistors by frequency dependent conductance analysis," *Appl. Phys. Lett.*, vol. 94, pp. 223512–223512-3, 2009. DOI: 10.1063/1.3148830. 10, 12

[31] U. K. Mishra, L. Shen, T. E. Kazior, and W. Yi-Feng, "GaN-based RF power devices and amplifiers," *Proc. IEEE*, vol. 96, pp. 287–305, 2008. DOI: 10.1109/JPROC.2007.911060. 12

[32] M. Asif Khan, A. Bhattari, J. N. Kuznia, and D. T. Olson, "High electron mobility transistor based on a GaN-AlxGa1-xN heterojunction," *Appl. Phys. Lett.*, vol. 63, pp. 1214–5, 1993. DOI: 10.1063/1.109775. 12

[33] M. L. Schuette, A. Ketterson, B. Song, E. Beam, T. M. Chou, M. Pilla, H. Q. Tserng, X. Gao, S. Gui, P. J. Fay, H. G. Xing, and P. Saunier, "Gate-recessed integrated E/D GaN HEMT technology with fT/fmax > 300 GHz," *IEEE Electron Device Lett.*, vol. 34, pp. 741–3, 2013. DOI: 10.1109/LED.2013.2257657. 13, 14, 16

[34] N. Nidhi, S. Dasguota, J. Lu, J. S. Speck, and U. K. Mishra, "Scaled self-aligned N-polar GaN/AlGaN MIS-HEMTs with fT of 275 GHz," *IEEE Electron Device Lett.*, vol. 33, pp. 961–3, 2012. DOI: 10.1109/LED.2012.2194130. 13, 14, 16

[35] D. F. Brown, A. Williams, K. Shinohara, A. Kurdoghlian, I. Milosavljevic, P. Hashimoto, R. Grabar, S. Burnham, C. Butler, P. Willadsen, and M. Micovic, "W-band power performance of AlGaN/GaN DHFETs with regrown n+ GaN ohmic contacts by MBE," in *IEEE International Electron Devices Meeting (IEDM)*, Washington, DC, pp. 19.3.1–19.3.4, 2011. DOI: 10.1109/IEDM.2011.6131584. 13

[36] G. Meneghesso, M. Meneghini, F. Medjdoub, Y. Tagro, B. Grimbert, D. Ducatteau, N. Rolland, R. Silvestri, and E. Zanoni, "First reliability demonstration of sub-200-nm AlN/GaN-on-silicon double-heterostructure HEMTs for Ka-band applications," *IEEE Trans. Device Mater. Reliab.*, vol. 13, pp. 480–8, 2013. DOI: 10.1109/TDMR.2013.2276425. 13, 14

[37] C. Prashant and M. Umesh, "High electron mobility transistors (HEMTs)" in *RF and Microwave Passive and Active Technologies*, M. Golio and J. Golio, Eds., ed Boca Raton, FL: CRC Press, pp. 21.1–21.31, 2007. 13, 15

[38] P. Saunier, M. L. Schuette, C. Tso-Min, A. Hua-Quen, A. Ketterson, E. Beam, M. Pilla, and G. Xiang, "InAlN barrier scaled devices for very high fT and for low-voltage RF applications," *IEEE Trans. Electron Devices*, vol. 60, pp. 3099–3104, 2013. DOI: 10.1109/TED.2013.2277772. 13, 14

[39] M. H. Wong, S. Keller, S. D. Nidhi, D. J. Denninghoff, S. Kolluri, D. F. Brown, J. Lu, N. A. Fichtenbaum, E. Ahmadi, U. Singisetti, A. Chini, S. P. DenBaars, J. S. Speck, and U. K. Mishra, "N-polar GaN epitaxy and high electron mobility transistors" *Semicond. Sci. Technol.*, vol. 28, p. 074009, 2013. DOI: 10.1088/0268-1242/28/7/074009. 13, 14, 16, 19

[40] H. Tongde, L. Z. Jun, Z. Zueliang, M. Jun, L. Xing, and L. Kei May, "DC and RF performance on gate-last AlN/GaN MOSHEMTs on Si with regrown source/drain," *IEEE Trans. Electron Devices*, vol. 60, pp. 3019–24, 2013. DOI: 10.1109/TED.2013.2274656. 14, 16

[41] X. Lu, J. Ma, P. Xu, H. Jiang, and K. M. Lau, "High performance self-aligned AlN-GaN MISHEMT with in situ SiNx gate dielectric and regrown source-drain," in *International Conference on Compound Semiconductor Manufacturing Technology*, Denver, Co, pp. 201–3, 2014. 14, 16

[42] C. Yong, C. Zhiqun, Y. Zhenchuan, T. Chak Wah, K. M. Lau, and K. J. Chen, "High temperature operation of AlGaN/GaN HEMTs direct-coupled FET logic (DCFL) integrated circuits," *IEEE Electron Device Lett.*, vol. 28, pp. 328–31, 2007. DOI: 10.1109/LED.2007.895391. 14

[43] L. Xing, M. Jun, H. Jiang, C. Liu, P. Xu, and K. M. Lau, "Fabrication and Characterization of Gate-Last Self-Aligned AlN/GaN MISHEMTs With In Situ SiN$_x$ Gate Dielectric," *IEEE Trans. Electron Devices*, 2015. in press. DOI: 10.1109/TED.2015.2421031. 14, 16, 17, 20, 21

[44] G. Jia, L. Gouwang, F. Faria, C. Yu, W. Ronghua, J. Verma, G. Xiang, G. Shiping, E. Beam, A. Ketterson, M. Schuette, P. Saunier, M. Wistey, D. Jena, and X. Huili, "MBE-regrown ohmics in InAlN HEMTs with a regrowth interface resistance of 0.05 ohm.mm," *IEEE Electron Device Lett.*, vol. 33, pp. 525–7, 2012. DOI: 10.1109/LED.2012.2186116. 14

[45] A. L. Corrion, K. Shinohara, D. Regan, Y. Tang, D. Brown, J. F. Robinson, H. H. Fung, A. Schmitz, D. Le, S. J. Kim, T. C. Oh, and M. Micovic, "High-speed 501-stage DCFL GaN ring oscillator circuits," *IEEE Electron Device Lett.*, vol. 34, pp. 846–8, 2013. DOI: 10.1109/LED.2013.2264796. 16

[46] D. J. Denninghoff, S. Dasgupta, J. Lu, S. Keller, and U. K. Mishra, "Design of high-aspect-ratio t-gates on N-polar GaN/AlGaN MIS-HEMTs for high fmax," *IEEE Electron Device Lett.*, vol. 33, pp. 785–7, 2012. DOI: 10.1109/LED.2012.2191134. 16, 18

[47] Nidhi, S. Dasgupta, D. F. Brown, U. Singisetti, S. Keller, J. S. Speck, and U. K. Mishra, "Self-aligned technology for N-polar GaN/Al(Ga)N MIS-HEMTs," *IEEE Electron Device Lett.*, vol. 32, pp. 33–5, 2011. DOI: 10.1109/LED.2010.2086427. 16

[48] J. Everts, J. Das, J. Van dan Keybus, M. Germain, and J. Driesen, "GaN-based power transistors for future power electronic converters," in *Proceedings of Young Researchers Symposium 2010 on Smart Sustainable Power Delivery*, 2010. 20

[49] S. Bandyopadhyay, B. Neidorff, D. Freeman, and A. P. Chandrakasan, "90.6% efficient 11MHz 22W LED driver using GaN FETs and burst-mode controller with 0.96 power factor," in *IEEE International Solid-State Circuits Conference Digest of Technical Papers (ISSCC)*, San Francisco, CA, pp. 368–369, 2013. DOI: 10.1109/ISSCC.2013.6487773. 20

[50] W. Saito, T. Nitta, Y. Kakiuchi, Y. Saito, K. Tsuda, I. Omura, and M. Yamaguchi, "A 120-W boost converter operation using a high-voltage GaN-HEMT," *IEEE Electron Device Lett.*, vol. 29, pp. 8–10, 2008. DOI: 10.1109/LED.2007.910796.

[51] W. Chen, K.-Y. Wong, and K. J. Chen, "Monolithic integration of lateral field-effect rectifier with normally-off HEMT for GaN-on-Si switch-mode power supply converters," in *IEEE International Electron Devices Meeting, (IEDM)*, San Francisco, CA, pp. 1–4, 2008. DOI: 10.1109/IEDM.2008.4796635.

[52] A. Lidow, J. Strydom, M. de Rooij, and Y. Ma, *GaN transistors for efficient power conversion*: John Wiley & Sons Incorporated, 2014. DOI: 10.1002/9781118844779. 20

[53] J. Ma, X. Zhu, T. Huang, L. Chao, and K. M. Lau, "Improved buffer rersistivity for GaN-based HEMTs using a medium-temperature and low -pressure GaN insertion layer," in *17th International Conference on Metal Organic Vapor Phase Epitaxy*, Lausanne, Switzerland, 2014. 20

[54] International_Rectifier. *SMPS MOSFET IRF830APbF*. Available: http://datasheet.octopart.com/IRF830APBF-Vishay-datasheet-14723025.pdf 21

[55] W. Saito, T. Nitta, Y. Kakiuchi, Y. Saito, K. Tsuda, I. Omura, and M. Yamaguchi, "Suppression of dynamic on-resistance increase and gate charge measurements in high-voltage GaN-HEMTs with optimized field-plate structure," *IEEE Trans. Electron Devices*, vol. 54, pp. 1825–1830, 2007. DOI: 10.1109/TED.2007.901150. 22

[56] Panasonic. (2015). *GaN power devices*. Available: `http://www.semicon.panasonic.co.jp/en/products/powerics/ganpower/` 22

[57] Z. Li, J. Waldron, T. Detchprohm, C. Wetzel, R. F. Karlicek, and T. P. Chow, "Monolithic integration of light-emitting diodes and power metal-oxide-semiconductor channel high-electron-mobility transistors for light-emitting power integrated circuits in GaN on sapphire substrate," *Appl. Phys. Lett.,* vol. 102, p. 192107, 2013. DOI: 10.1063/1.4807125. 22, 26

[58] Z. J. Liu, H. Tongde, J. Ma, C. Liu, and K. M. Lau, "Monolithic Integration of AlGaN/GaN HEMT on LED by MOCVD," *IEEE Electron Device Lett.*, vol. 35, pp. 330–332, 2014. DOI: 10.1109/LED.2014.2300897. 22, 26

[59] Z. Liu, J. Ma, H. Tongde, C. Liu, and K. M. Lau, "Selective epitaxial growth of monolithically integrated GaN-based light emitting diodes with AlGaN/GaN driving transistors," *Appl. Phys. Lett.*, vol. 104, 2014. DOI: 10.1063/1.4867235. 22, 24, 26

[60] K. M. Lau, K. M. Wong, X. Zou, and P. Chen, "Performance improvement of GaN-based light-emitting diodes grown on patterned Si substrate transferred to copper," *Opt. Express,* vol. 19, pp. A956-A961, 2011. DOI: 10.1364/OE.19.00A956. 26

[61] T. Mimura, S. Hiyamizu, T. Fujii, and K. Nanbu, "A new field-effect transistor with selectively doped GaAs/n-AlxGa1-xAs heterojunctions," *Jpn. J. Appl. Phys.*, vol. 19, p. L225, 1980. DOI: 10.1143/JJAP.19.L225. 31, 32, 34

[62] N. Moll, M. R. Hueschen, and A. Fischer-Colbrie, "Pulse-doped AlGaAs/InGaAs pseudomorphic MODFETs," *IEEE Trans. Electron Devices*, vol. 35, pp. 879–886, 1988. DOI: 10.1109/16.3339. 31, 32

[63] G. Munns, M. E. Sherwin, T. Brock, G. I. Haddad, Y. Kwon, G. Ng, and D. Pavlidis, "InAlAs/InGaAs/InP sub-micron HEMTs grown by CBE," *J. Cryst. Growth*, vol. 120, pp. 184–188, 1992. DOI: 10.1016/0022-0248(92)90388-Y. 31

[64] J. B. Boos, W. Kruppa, B. R. Bennett, D. Park, S. W. Kirchoefer, R. Bass, and H. B. Dietrich, "AlSb/InAs HEMT's for low-voltage, high-speed applications," *IEEE Trans. Electron Devices*, vol. 45, pp. 1869–1875, 1998. DOI: 10.1109/16.711349. 31

[65] S. Datta, T. Ashley, J. Brask, L. Buckle, M. Doczy, M. Emeny, D. Hayes, K. Hilton, R. Jefferies, and T. Martin, "85 nm gate length enhancement and depletion mode InSb quantum well transistors for ultra high speed and very low power digital logic applications," *IEDM Tech. Dig.*, pp. 763–766, 2005. DOI: 10.1109/IEDM.2005.1609466. 31

[66] H. L. Störmer and W. T. Tsang, "Two-dimensional hole gas at a semiconductor heterojunction interface," *Appl. Phys. Lett.*, vol. 36, pp. 685–687, 1980. DOI: 10.1063/1.91624. 31

[67] I. J. Fritz, L. R. Dawson, T. J. Drummond, J. E. Schirber, and R. M. Biefeld, "Hall-effect measurements in p-type InGaAs/GaAs strained-layer superlattices," *Appl. Phys. Lett.,* vol. 48, pp. 139–141, 1986. DOI: 10.1063/1.96976. 31

[68] M. D. Feuer, Y. He, D. M. Tennant, S. C. Shunk, K. F. Brown-Goebeler, R. E. Behringer, and T. Y. Chang, "InP-based HIGFETs for complementary circuits," *IEEE Trans. Electron Devices,* vol. 36, p. 2616, 1989. DOI: 10.1109/16.43729. 31

[69] B. R. Bennett, M. G. Ancona, J. B. Boos, C. B. Canedy, and S. A. Khan, "Strained GaSb/AlAsSb quantum wells for p-channel field-effect transistors," *J. Cryst. Growth,* vol. 311, pp. 47–53, 12/15/ 2008. DOI: 10.1016/j.jcrysgro.2008.10.025. 31

[70] B. R. Bennett, M. G. Ancona, J. B. Boos, and B. V. Shanabrook, "Mobility enhancement in strained p-InGaSb quantum wells," *Appl. Phys. Lett.,* vol. 91, 2007. DOI: 10.1063/1.2762279. 31

[71] H. Sugiyama, T. Hoshi, H. Yokoyama, and H. Matsuzaki, "Metal-organic vapor-phase epitaxy growth of InP-based HEMT structures with InGaAs/InAs composite channel," in *International Conference on Indium Phosphide and Related Materials (IPRM)*, Santa Barbara, CA, pp. 245–248, 2012. DOI: 10.1109/ICIPRM.2012.6403369. 32

[72] E. Y. Chang, C. I. Kuo, H. T. Hsu, C. Y. Chiang, and y. Miyamoto, "InAs thin-channel high-electron-mobility transistors with very high current-gain cutoff frequency for emerging submillimeter-wave applications," *Appl. Phys. Express*, vol. 6, p. 034001, 2013. DOI: 10.7567/APEX.6.034001. 32, 35

[73] E. F. Schubert, "Delta doping of III–V compound semiconductors: fundamentals and device applications," *J. Vac. Sci. Technol., A*, vol. 8, pp. 2980–2996, 1990. DOI: 10.1116/1.576617. 32

[74] T. Akazaki, T. Enoki, K. Arai, and Y. Ishii, "Improving the characteristics of an InAlAsInGaAs inverted HEMT by inserting an InAs layer into the InGaAs channel," *Solid State Electron.*, vol. 38, pp. 997–1000, 1995. DOI: 10.1016/0038-1101(95)98667-R. 32

[75] M. Egard, L. Ohlsson, B. M. Borg, F. Lenrick, R. Wallenberg, L. E. Wernersson, and E. Lind, "High transconductance self-aligned gate-last surface channel In0.53Ga0.47 As MOSFET," in *IEEE International Electron Devices Meeting (IEDM)*, Washington, DC, pp. 13.2.1–13.2.4, 2011. DOI: 10.1109/IEDM.2011.6131544. 32, 39

[76] Q. Li, X. Zhou, C. Tang, and K. M. Lau, "High-performance inverted MOSHEMTs on a GaAs Substrate With Regrown Source/Drain by MOCVD," *IEEE Electron Device Lett.,* vol. 33, pp. 1246–1248, 2012. DOI: 10.1109/LED.2012.2204431. 32, 38, 39

[77] C. M. Wolfe, G. E. Stillman, and W. T. Lindley, "Electron mobility in high-purity GaAs," *J. Appl. Phys.*, vol. 41, pp. 3088–3091, 1970. DOI: 10.1063/1.1659368. 34

[78] W. Walukiewicz, H. E. Ruda, J. Lagowski, and H. C. Gatos, "Electron mobility in modulation-doped heterostructures," *Phys. Rev. B*, vol. 30, pp. 4571–4582, 10/15/ 1984. DOI: 10.1103/PhysRevB.30.4571. 34

[79] J. A. del Alamo, "The high electron mobility transistor at 30: impressive accomplishments and exciting prospects" in *International Conference on Compound Semiconductor Manufacturing Technology*, pp. 17–22, 2011. 34, 35

[80] H. Matsuzaki, T. Maruyama, T. Kosugi, H. Takahashi, M. Tokumitsu, and T. Enoki, "Laterally scaled down tiered-edge ohmic structure of InP-based HEMTs for 2-S/mmg/sub m/ and 500-GHz f/sub T," in *IEEE InternationalElectron Devices Meeting*, Washington, DC, pp. 775–778, 2005. DOI: 10.1109/IEDM.2005.1609469. 34

[81] T.-W. Kim, D.-H. Kim, and J. A. Del Alamo, "60 nm self-aligned-gate InGaAs HEMTs with record high-frequency characteristics," in *IEEE International Electron Devices Meeting (IEDM)*, San Francisco, CA, pp. 30.7.1–30.7.4, 2010. DOI: 10.1109/IEDM.2010.5703454. 35

[82] R. Lai, X. B. Mei, W. R. Deal, W. Yoshida, Y. M. Kim, P. H. Liu, J. Lee, J. Uyeda, V. Radisic, M. Lange, T. Gaier, L. Samoska, and A. Fung, "Sub 50 nm InP HEMT device with Fmax greater than 1 THz," in *IEEE International Electron Devices Meeting*, Washington, DC, pp. 609–611, 2007. DOI: 10.1109/IEDM.2007.4419013. 35

[83] D. H. Kim, J. A. del Alamo, P. Chen, W. Ha, M. Urteaga, and B. Brar, "50-nm E-mode In0.7Ga0.3 As PHEMTs on 100-mm InP substrate with fmax > 1 THz," in *IEEE International Electron Devices Meeting (IEDM)*, San Francisco, CA, pp. 30.6.1–30.6.4, 2010. DOI: 10.1109/IEDM.2010.5703453. 35

[84] A. Leuther, S. Koch, A. Tessmann, I. Kallfass, T. Merkle, H. Massler, R. Loesch, M. Schlechtweg, S. Saito, and O. Ambacher, "20 nm metamorphic HEMT with 660 GHz FT," in *23rd International Conference on Indium Phosphide and Related Materials Compound Semiconductor Week (CSW/IPRM)*, Berlin, pp. 1–4, 2011. 35

[85] D.-H. Kim and J. A. Del Alamo, "30-nm InAs PHEMTs with ft = 644 GHz and fmax = 681 GHz," *IEEE Electron Device Lett.*, vol. 31, pp. 806–808, 2010. DOI: 10.1109/LED.2010.2051133. 35

[86] E. Pop, "Energy dissipation and transport in nanoscale devices," *Nano Res.*, vol. 3, pp. 147–169, 2010. DOI: 10.1007/s12274-010-1019-z. 36

[87] J. A. del Alamo and D.-H. Kim, "Beyond CMOS: logic suitability of InGaAs HEMTs," in *IEEE 19th International Conference on Indium Phosphide & Related Materials, IPRM'07*, Matsue, pp. 51–54, 2007. DOI: 10.1109/ICIPRM.2007.381120. 37

[88] D.-H. Kim and J. Del Alamo, "Logic performance of 40 nm InAs HEMTs," in *IEEE International Electron Devices Meeting,(IEDM)*, pp. 629–632, 2007. DOI: 10.1109/IEDM.2007.4419018. 37

[89] T.-W. Kim, D.-H. Kim, and J. A. del Alamo, "30 nm In0.7Ga0.3As inverted-type HEMTs with reduced gate leakage current for logic applications," in *IEEE International Electron Devices Meeting (IEDM)*, Baltimore, MD, pp. 1–4, 2009. DOI: 10.1109/IEDM.2009.5424317. 37

[90] P. Ye, G. Wilk, J. Kwo, B. Yang, H.-J. Gossmann, M. Frei, S. Chu, J. Mannaerts, M. Sergent, and M. Hong, "GaAs MOSFET with oxide gate dielectric grown by atomic layer deposition," *IEEE Electron Device Lett.*, vol. 24, pp. 209–211, 2003. DOI: 10.1109/LED.2003.812144. 37

[91] M. Hong, J. Kwo, A. Kortan, J. Mannaerts, and A. Sergent, "Epitaxial cubic gadolinium oxide as a dielectric for gallium arsenide passivation," *Science*, vol. 283, pp. 1897–1900, 1999. DOI: 10.1126/science.283.5409.1897. 38

[92] J. Kwo, D. Murphy, M. Hong, R. Opila, J. Mannaerts, A. Sergent, and R. Masaitis, "Passivation of GaAs using (Ga2O3) 1-x (Gd2O3) x, $0 \leq x \leq 1.0$ films," *Appl. Phys. Lett.*, vol. 75, pp. 1116–1118, 1999. DOI: 10.1063/1.124614.

[93] K. Rajagopalan, R. Droopad, J. Abrokwah, P. Zurcher, P. Fejes, and M. Passlack, "1-μm enhancement mode GaAs n-channel MOSFETs with transconductance exceeding 250 mS/mm," *IEEE Electron Device Lett.*, vol. 28, pp. 100–102, 2007. DOI: 10.1109/LED.2006.889502. 38

[94] Y. Xuan, T. Shen, M. Xu, Y. Wu, and P. Ye, "High-performance surface channel In-rich In 0.75 Ga 0.25 As MOSFETs with ALD high-k as gate dielectric," in *IEEE International Electron Devices Meeting*, pp. 1–4, 2008. DOI: 10.1109/IEDM.2008.4796697. 38

[95] M. M. Frank, G. D. Wilk, D. Starodub, T. Gustafsson, E. Garfunkel, Y. J. Chabal, J. Grazul, and D. A. Muller, "HfO 2 and Al 2 O 3 gate dielectrics on GaAs grown by atomic layer deposition," *Appl. Phys. Lett.*, vol. 86, pp. 152904–152904-3, 2005. DOI: 10.1063/1.1899745. 38

[96] N. Goel, D. Heh, S. Koveshnikov, I. Ok, S. Oktyabrsky, V. Tokranov, R. Kambhampatic, M. Yakimov, Y. Sun, and P. Pianetta, "Addressing the gate stack challenge for high mobility In x Ga 1-x As channels for NFETs," in *IEEE International Electron Devices Meeting*, San Francisco, CA, pp. 1–4, 2008. DOI: 10.1109/IEDM.2008.4796695. 38

[97] J. Huang, N. Goel, H. Zhao, C. Kang, K. Min, G. Bersuker, S. Oktyabrsky, C. Gaspe, M. Santos, and P. Majhi, "InGaAs MOSFET performance and reliability improvement by simultaneous reduction of oxide and interface charge in ALD (La) AlOx/ZrO 2 gate stack," in *IEEE International Electron Devices Meeting (IEDM)*, Baltimore, MD, pp. 1–4, 2009. DOI: 10.1109/IEDM.2009.5424357. 38

[98] W. Li, X. Wang, Y. Liu, S. Shim, and T. Ma, "Demonstration of unpinned GaAs surface and surface inversion with gate dielectric made of Si 3 N 4," *Appl. Phys. Lett.*, vol. 90, pp. 193503–193503-3, 2007. DOI: 10.1063/1.2737374. 38

[99] Y. Cao, X. Li, J. Zhang, P. Fay, T. Kosel, and D. Hall, "Microwave performance of GaAs MOSFET with wet thermally oxidized InAlP gate dielectric," *IEEE Electron Device Lett.*, vol. 27, pp. 317–319, 2006. DOI: 10.1109/LED.2006.872898. 38

[100] M. Radosavljevic, B. Chu-Kung, S. Corcoran, G. Dewey, M. Hudait, J. Fastenau, J. Kavalieros, W. Liu, D. Lubyshev, and M. Metz, "Advanced high-k gate dielectric for high-performance short-channel In 0.7 Ga 0.3 As quantum well field effect transistors on silicon substrate for low power logic applications," in *IEEE International Electron Devices Meeting (IEDM)*, Baltimore, MD, pp. 1–4, 2009. DOI: 10.1109/IEDM.2009.5424361. 38

[101] I. Ok, H. Kim, M. Zhang, T. Lee, F. Zhu, L. Yu, S. Koveshnikov, W. Tsai, V. Tokranov, and M. Yakimov, "Self-aligned n-and p-channel GaAs MOSFETs on undoped and p-type substrates using HfO2 and silicon interface passivation layer," in *International Conference on Electron Devices Meeting*, San Francisco, CA, pp. 1–4, 2006. DOI: 10.1109/IEDM.2006.346742. 38

[102] M. Zhu, C.-H. Tung, and Y.-C. Yeo, "Aluminum oxynitride interfacial passivation layer for high-permittivity gate dielectric stack on gallium arsenide," *Appl. Phys. Lett.*, vol. 89, p. 202903, 2006. DOI: 10.1063/1.2388246. 38

[103] J. Lin, S. Lee, H.-J. Oh, W. Yang, G. Lo, D. Kwong, and D. Chi, "Plasma PH 3-passivated high mobility inversion InGaAs MOSFET fabricated with self-aligned gate-first process and HfO 2/TaN gate stack," in *IEEE International Electron Devices Meeting (IEDM)*, pp. 1–4, 2008. DOI: 10.1109/IEDM.2008.4796705. 38

[104] H. Oh, J. Lin, S. Suleiman, G. Lo, D. Kwong, D. Chi, and S. Lee, "Thermally robust phosphorous nitride interface passivation for InGaAs self-aligned gate-first n-MOSFET integrated with high-k dielectric," in *IEEE International Electron Devices Meeting (IEDM)*, Baltimore, MD, pp. 1–4, 2009. DOI: 10.1109/IEDM.2009.5424354. 38

[105] C. Hock-Chun Chin, M. Zhu, X. Liu, H.-K. Lee, L. Shi, L.-S. Tan, and Y.-C. Yeo, "Silane–ammonia surface passivation for gallium arsenide surface-channel

n-MOSFETs," *IEEE Electron Device Lett.*, vol. 30, pp. 110–112, 2009. DOI: 10.1109/LED.2008.2010831. 38

[106] Y. Wu, M. Xu, R. Wang, O. Koybasi, and P. Ye, "High performance deep-submicron inversion-mode InGaAs MOSFETs with maximum G m exceeding 1.1 mS/μm: new HBr pretreatment and channel engineering," in *IEEE International Electron Devices Meeting (IEDM)*, Baltimore, MD, pp. 1–4, 2009. DOI: 10.1109/IEDM.2009.5424358. 38

[107] É. O'Connor, B. Brennan, V. Djara, K. Cherkaoui, S. Monaghan, S. Newcomb, R. Contreras, M. Milojevic, G. Hughes, and M. Pemble, "A systematic study of (NH4) 2S passivation (22%, 10%, 5%, or 1%) on the interface properties of the Al2O3/In0.53Ga0.47As/InP system for n-type and p-type In0.53Ga0.47As epitaxial layers," *J. Appl. Phys.*, vol. 109, p. 024101, 2011. DOI: 10.1063/1.3533959. 38

[108] T. Lin, H. Chiu, P. Chang, L. Tung, C. Chen, M. Hong, J. Kwo, W. Tsai, and Y. Wang, "High-performance self-aligned inversion-channel In0.53Ga0.47As metal-oxide-semiconductor field-effect-transistor with Al2O3/Ga2O3 (Gd2O3) as gate dielectrics," *Appl. Phys. Lett.*, vol. 93, p. 033516, 2008. DOI: 10.1063/1.2956393. 38

[109] A. Sonnet, R. Galatage, P. Hurley, E. Pelucchi, K. Thomas, A. Gocalinska, J. Huang, N. Goel, G. Bersuker, and W. Kirk, "On the calculation of effective electric field in In0.53Ga0.47As surface channel metal-oxide-semiconductor field-effect-transistors," *Appl. Phys. Lett.*, vol. 98, p. 3501, 2011. DOI: 10.1063/1.3588255. 38

[110] U. Singisetti, M. A. Wistey, G. J. Burek, A. K. Baraskar, J. Cagnon, B. Thibeault, A. C. Gossard, S. Stemmer, M. J. Rodwell, and E. Kim, "Enhancement mode In0.53Ga0.47As MOSFET with self-aligned epitaxial source/drain regrowth," in *IEEE International Conference on Indium Phosphide & Related Materials*, Newport Beach, CA, pp. 120–123, 2009. DOI: 10.1109/ICIPRM.2009.5012456. 38, 39

[111] H. Tsuchiya, A. Maenaka, T. Mori, and Y. Azuma, "Role of carrier transport in source and drain electrodes of high-mobility MOSFETs," *IEEE Electron Device Lett.*, vol. 31, pp. 365–367, 2010. DOI: 10.1109/LED.2010.2040024. 39

[112] ITRS. (2012). *International technology roadmap for semiconductors* Available: http://www.itrs.net 39

[113] U. Singisetti, M. A. Wistey, G. J. Burek, A. K. Baraskar, B. J. Thibeault, A. C. Gossard, M. J. Rodwell, B. Shin, E. J. Kim, and P. C. McIntyre, "Channel mosfets with self-aligned inas source/drain formed by mee regrowth," *IEEE Electron Device Lett.*, vol. 30, pp. 1128–1130, 2009. DOI: 10.1109/LED.2009.2031304. 39

[114] S. Lee, C.-Y. Huang, D. Cohen-Elias, B. J. Thibeault, W. Mitchell, V. Chobpattana, S. Stemmer, A. C. Gossard, and M. J. Rodwell, "Highly scalable raised source/drain InAs

quantum well MOSFETs exhibiting ION= 482 μA/μm at IOFF= 100 nA/μm and VDD= 0.5 V," *IEEE Electron Device Lett.*, vol. 35, pp. 621–3, 2014. 39

[115] S. Lee, J. J. Law, A. D. Carter, B. J. Thibeault, W. Mitchell, V. Chobpattana, S. Kramer, S. Stemmer, A. C. Gossard, and M. J. Rodwell, "Substitutional-gate MOSFETs with composite channels and self-aligned MBE source–drain regrowth," *IEEE Electron Device Lett.*, vol. 33, pp. 1553–1555, 2012. DOI: 10.1109/LED.2012.2215572. 39

[116] X. Zhou, Q. Li, C. W. Tang, and K. M. Lau, "Inverted-type InGaAs metal–oxide– semiconductor high-electron-mobility transistor on Si substrate with maximum drain current exceeding 2 A/mm," *Appl. Phys. Express*, vol. 5, p. 104201, 2012. DOI: 10.1143/APEX.5.104201. 39

[117] X. Zhou, Q. Li, C. W. Tang, and K. M. Lau, "30nm enhancement-mode In 0.53 Ga 0.47 As MOSFETs on Si substrates grown by MOCVD exhibiting high transconductance and low on-resistance," in *2012 IEEE International Electron Devices Meeting (IEDM)*, San Francisco, CA, pp. 32.5.1–32.5.4, 2012. DOI: 10.1109/IEDM.2012.6479153.

[118] X. Zhou, Q. Li, C. W. Tang, and K. M. Lau, "30-nm Inverted MOSHEMTs on Si sub- strate grown by MOCVD with regrown source/drain," *IEEE Electron Device Lett.*, vol. 33, pp. 1384–1386, 2012. DOI: 10.1109/LED.2012.2210383. 39

[119] S. Kim, M. Yokoyama, N. Taoka, R. Iida, S. Lee, R. Nakane, Y. Urabe, N. Miyata, T. Yasuda, and H. Yamada, "Self-aligned metal source/drain InxGa1-xAs n-metal–oxide– semiconductor field-effect transistors using Ni–InGaAs alloy," *Appl. Phys. Express*, vol. 4, p. 024201, 2011. DOI: 10.1143/APEX.4.024201. 39

[120] S. Kim, M. Yokoyama, R. Nakane, O. Ichikawa, T. Osada, M. Hata, M. Take- naka, and S. Takagi, "High performance extremely-thin body InAs-on-insulator MOS- FETs on Si with Ni-InGaAs metal S/D by contact resistance reduction technology," in *Symposium on VLSI Technology (VLSIT)*, Kyoto, Japan, pp. T52-T53, 2013. DOI: 10.1109/TED.2013.2279363. 39, 40

[121] M. Radosavljevic, G. Dewey, D. Basu, J. Boardman, B. Chu-Kung, J. Fastenau, S. Kabehie, J. Kavalieros, V. Le, and W. Liu, "Electrostatics improvement in 3-D tri- gate over ultra-thin body planar InGaAs quantum well field effect transistors with high- K gate dielectric and scaled gate-to-drain/gate-to-source separation," in *IEEE Interna- tional Electron Devices Meeting (IEDM)*, Washington, DC, pp. 33.1.1–33.1.4, 2011. DOI: 10.1109/IEDM.2011.6131661. 40

[122] J. Gu, Y. Liu, Y. Wu, R. Colby, R. G. Gordon, and P. D. Ye, "First experimental demon- stration of gate-all-around III–V MOSFETs by top-down approach," in *IEEE Interna- tional Electron Devices Meeting (IEDM)*, Washington, DC, pp. 33.2.1–33.2.4, 2011. 40

[123] X. Miao and X. Li, "Scalable monolithically grown AlGaAs–GaAs planar nanowire high-electron-mobility transistor," *IEEE Electron Device Lett.*, vol. 32, pp. 1227–1229, 2011. DOI: 10.1109/LED.2011.2160248. 40

[124] K. Tomioka, M. Yoshimura, and T. Fukui, Vertical In 0.7 Ga 0.3 As nanowire surrounding-gate transistors with high-k gate dielectric on Si substrate," in *IEEE International Electron Devices Meeting (IEDM)*, Washington, DC, pp. 33.3.1–33.3.4, 2011. DOI: 10.1109/IEDM.2011.6131663. 40

[125] K. Tomioka, M. Yoshimura, and T. Fukui, "A III-V nanowire channel on silicon for high-performance vertical transistors," *Nature*, vol. 488, pp. 189–192, 2012. DOI: 10.1038/nature11293. 40

[126] M. Yokoyama, H. Takagi, T. Yasuda, H. Yamada, N. Fukuhara, M. Hata, M. Sugiyama, Y. Nakano, M. Takenaka, and S. Takagi, "Ultrathin body InGaAs-on-insulator metal–oxide–semiconductor field-effect transistors with InP passivation layers on Si substrates fabricated by direct wafer bonding," *Appl. Phys. Express*, vol. 4, p. 054202, 2011. DOI: 10.1143/APEX.4.054202. 41

[127] M. Yokoyama, S. Kim, R. Zhang, N. Taoka, Y. Urabe, T. Maeda, H. Takagi, T. Yasuda, H. Yamada, and O. Ichikawa, "III–V/Ge high mobility channel integration of InGaAs n-channel and Ge p-channel metal–oxide–semiconductor field-effect transistors with self-aligned Ni-based metal source/drain using direct wafer bonding," *Appl. Phys. Express*, vol. 5, p. 076501, 2012. DOI: 10.1143/APEX.5.076501. 41

[128] H. Ko, K. Takei, R. Kapadia, S. Chuang, H. Fang, P. W. Leu, K. Ganapathi, E. Plis, H. S. Kim, and S.-Y. Chen, "Ultrathin compound semiconductor on insulator layers for high-performance nanoscale transistors," *Nature*, vol. 468, pp. 286–289, 2010. DOI: 10.1038/nature09541. 41

[129] J. Nah, H. Fang, C. Wang, K. Takei, M. H. Lee, E. Plis, S. Krishna, and A. Javey, "III–V complementary metal–oxide–semiconductor electronics on silicon substrates," *Nano Lett.*, vol. 12, pp. 3592–3595, 2012. DOI: 10.1021/nl301254z. 41

[130] L. Desplanque, S. El Kazzi, C. Coinon, S. Ziegler, B. Kunert, A. Beyer, K. Volz, W. Stolz, Y. Wang, and P. Ruterana, "Monolithic integration of high electron mobility InAs-based heterostructure on exact (001) Silicon using a GaSb/GaP accommodation layer," *Appl. Phys. Lett.*, vol. 101, p. 142111, 2012. DOI: 10.1063/1.4758292. 42, 43

[131] K. M. Lau, C. W. Tang, H. Li, and Z. Zhong, "AlInAs/GaInAs mHEMTs on silicon substrates grown by MOCVD," in *IEEE International Electron Devices Meeting*, San Francisco, CA, pp. 1–4, 2008. DOI: 10.1109/IEDM.2008.4796797. 42, 43

[132] W. Liu, D. Lubyshev, J. Fastenau, Y. Wu, M. Bulsara, E. Fitzgerald, M. Urteaga, W. Ha, J. Bergman, and B. Brar, "Monolithic integration of InP-based transistors on Si substrates using MBE," *J. Cryst. Growth*, vol. 311, pp. 1979–1983, 2009. DOI: 10.1016/j.jcrysgro.2008.10.061. 42, 43

[133] S. Datta, G. Dewey, J. Fastenau, M. Hudait, D. Loubychev, W. Liu, M. Radosavljevic, W. Rachmady, and R. Chau, "Ultrahigh-speed 0.5 V supply voltage In 0.7 Ga 0.3 As quantum-well transistors on silicon substrate," *IEEE Electron Device Lett.*, vol. 28, pp. 685–687, 2007. DOI: 10.1109/LED.2007.902078. 42, 43

[134] M. K. Hudait, G. Dewey, S. Datta, J. M. Fastenau, J. Kavalieros, W. K. Liu, D. Lubyshev, R. Pillarisetty, W. Rachmady, M. Radosavljevic, T. Rakshit, and R. Chau, "Heterogeneous integration of enhancement mode In0.7Ga0.3As quantum well transistor on silicon substrate using thin (les 2 um) composite buffer architecture for high-speed and low-voltage (0.5 v) logic applications," in *IEEE International Electron Devices Meeting*, Washington, DC, pp. 625–628, 2007. 42

[135] N. Mukherjee, J. Boardman, B. Chu-Kung, G. Dewey, A. Eisenbach, J. Fastenau, J. Kavalieros, W. K. Liu, D. Lubyshev, M. Metz, K. Millard, M. Radosavljevic, T. Stewart, H. W. Then, P. Tolchinsky, and R. Chau, "MOVPE III-V material growth on silicon substrates and its comparison to MBE for future high performance and low power logic applications," in *IEEE International Electron Devices Meeting (IEDM)*, Washington, DC, pp. 35.1.1–35.1.4, 2011. DOI: 10.1109/IEDM.2011.6131675.

[136] Q. Li, C. W. Tang, and K. M. Lau, "Growth of ultra-high mobility In0.52Al0.48As/InxGa1- xAs (x≥ 53%) quantum wells on Si substrates using InP/GaAs buffers by metalorganic chemical vapor deposition," *Appl. Phys. Express*, vol. 7, p. 045502, 2014. DOI: 10.7567/APEX.7.045502. 42

[137] Y. Lin, H. Yamaguchi, E. Chang, Y. Hsieh, M. Ueki, Y. Hirayama, and C. Chang, "Growth of very-high-mobility AlGaSb/InAs high-electron-mobility transistor structure on si substrate for high speed electronic applications," *Appl. Phys. Lett.*, vol. 90, pp. 023509–023509-3, 2007. DOI: 10.1063/1.2431567. 42

[138] T. Langdo, C. Leitz, M. Currie, E. Fitzgerald, A. Lochtefeld, and D. Antoniadis, "High quality Ge on Si by epitaxial necking," *Appl. Phys. Lett.*, vol. 76, pp. 3700–3702, 2000. DOI: 10.1063/1.126754. 42

[139] J.-S. Park, J. Bai, M. Curtin, B. Adekore, M. Carroll, and A. Lochtefeld, "Defect reduction of selective Ge epitaxy in trenches on Si (001) substrates using aspect ratio trapping," *Appl. Phys. Lett.*, vol. 90, pp. 052113–052113-3, 2007. DOI: 10.1063/1.2435603. 42

[140] J. Li, J. Bai, J.-S. Park, B. Adekore, K. Fox, M. Carroll, A. Lochtefeld, and Z. Shellen-barger, "Defect reduction of GaAs epitaxy on Si (001) using selective aspect ratio trapping," *Appl. Phys. Lett.*, vol. 91, pp. 021114–021114-3, 2007. DOI: 10.1063/1.2756165. 42

[141] J. Li, J. Bai, J. M. Hydrick, J. G. Fiorenza, C. Major, M. Carroll, Z. Shellenbarger, and A. Lochtefeld, "Thin film InP epitaxy on Si (001) using selective aspect ratio trapping," *ECS Trans.*, vol. 18, pp. 887–894, 2009. DOI: 10.1149/1.3096551. 42

[142] C. Merckling, N. Waldron, S. Jiang, W. Guo, O. Richard, B. Douhard, A. Moussa, D. Vanhaeren, H. Bender, and N. Collaert, "Selective area growth of InP in shallow trench isolation on large scale Si (001) wafer using defect confinement technique," *J. Appl. Phys.*, vol. 114, p. 033708, 2013. DOI: 10.1063/1.4815959. 44

[143] M. Paladugu, C. Merckling, R. Loo, O. Richard, H. Bender, J. Dekoster, W. Vander-vorst, M. Caymax, and M. Heyns, "Site selective integration of III-V materials on Si for nanoscale logic and photonic devices," *Cryst. Growth Des.*, vol. 12, pp. 4696–4702, 2012. DOI: 10.1021/cg300779v. 44

[144] C. Merckling, N. Waldron, S. Jiang, W. Guo, N. Collaert, M. Caymax, E. Vancoille, K. Barla, A. Thean, and M. Heyns, "Heteroepitaxy of InP on Si (001) by selective-area metal organic vapor-phase epitaxy in sub-50 nm width trenches: the role of the nucle-ation layer and the recess engineering," *J. Appl. Phys.*, vol. 115, p. 023710, 2014. DOI: 10.1063/1.4862044. 44

[145] R. Cipro, T. Baron, M. Martin, J. Moeyaert, S. David, V. Gorbenko, F. Bassani, Y. Bogu-milowicz, J. P. Barnes, N. Rochat, V. Loup, C. Vizioz, N. Allouti, N. Chauvin, X. Y. Bao, Z. Ye, J. B. Pin, and E. Sanchez, "Low defect InGaAs quantum well selectively grown by metal organic chemical vapor deposition on Si(100) 300 mm wafers for next generation non planar devices," *Appl. Phys. Lett.*, vol. 104, p. 262103, 2014. DOI: 10.1063/1.4886404. 44

Authors' Biographies

ZHAOJUN LIU

Zhaojun Liu, Ph.D., is currently an assistant professor with Sun Yat-sen University-Carnegie Mellon University Joint Institute of Engineering (JIE) and Shunde International Joint Research Institute (JRI). He received the B.Eng. degree in Electrical Engineering and Automation from the Hebei University of Technology in 2003, the M.Phil. degree in Physical Electronics from Nankai University in 2007, and the Ph.D. degree in Electronic and Computer Engineering (ECE) from the Hong Kong University of Science and Technology (HKUST) in 2011. He was a research assistant with the State Key Laboratory on Advanced Displays and Optoelectronics Technologies (PSKL) at HKUST in Oct. 2006-Apr. 2007. He was a postdoctoral fellow with the Photonics Technology Center (PTC) at HKUST in Jan. 2011-July 2013. He was a visiting assistant professor with the ECE department at HKUST in Aug. 2013-Sept. 2014. He is an adjunct assistant professor with ECE department at HKUST since Sept. 2014 and a visiting professor with ECE department at Carnegie Mellon University since Dec. 2015. Dr. Liu is the core member of the "Pearl River Talent Plan" 5th Innovation Team of Guangdong Province. His research interests focused on (1) GaN LED micro-display, solid-state lighting and monolithic integration with GaN HEMTs; (2) Wearable Electronic & Photonic devices, and (3) 2D materials & devices.

TONGDE HUANG

Tongde Huang received his B.S. degree from Jilin University, Changchun in 2006, and M.S. degree from Shanghai Institute of Ceramics, Chinese Academy of Science, Shanghai in 2009. His Ph.D. degree was obtained at the Hong Kong University of Science and Technology in 2013. He joined the Microwave Electronic Laboratory in Chalmers University of Technology as a postdoc researcher from 2014. His research focuses on device design, modeling, and in-house MMIC. Current efforts include designing high-linearity GaN power amplifier, noise measurement, and modeling.

QIANG LI

Qiang Li, Ph.D., is currently a research assistant professor with the Department of Electronic and Computer Engineering at the Hong Kong University of Science and Technology (HKUST). He received his B.S. degree in Microelectronics from Peking University, Beijing, China in 2009 and a Ph.D. in Electronic and Computer Engineering from HKUST in 2014. His research focuses on compound semiconductor materials and devices.

XING LU

Xing Lu, Ph.D., is currently an Assistant Professor with the State Key Laboratory of Electrical Insulation and Power Equipment, School of Electrical Engineering, Xi'an Jiaotong University, Xi'an, China. He received a B.S. degree in Microelectronics from Fudan University, Shanghai, China in 2010, and a Ph.D. degree in Electronic and Computer Engineering from the Hong Kong University of Science and Technology, Hong Kong in 2014. His research focuses on design and implementation of compound semiconductor devices.

XINBO ZOU

Xinbo Zou, Ph.D., is currently a research assistant professor with Department of Electronic and Computer Engineering and also a Junior Fellow with Institute for Advanced Study (IAS), HKUST. He received his B.Eng. degree in Electronic Science and Technology from Beijing University of Posts and Telecommunications (BUPT) in 2007 and his Ph.D. from the Hong Kong University of Science and Technology (HKUST) in 2013 for developing high-performance green and yellow LEDs on Si substrates. His research interests focused on (1) development and characterization of novel GaN-based nanostructures grown on Si; and (2) MOCVD growth and fabrication of GaN-based devices.

Printed in the United States
by Baker & Taylor Publisher Services